the complete guide to understanding & caring for your cat

the complete guide to understanding & caring for your cat

Carole C. Wilbourn

The Cat Therapist

New York / London

www.sterlingpublishing.com

STERLING and the distinctive Sterling logo are registered trademarks of
Sterling Publishing Co., Inc.

Library of Congress Cataloging-in-Publication Data

Wilbourn, Carole, 1940–
The complete guide to understanding
and caring for your cat /
Carole C. Wilbourn.
p. cm.
Includes index.
ISBN-13 : 978-1-4027-0636-3
ISBN-10 : 1-4027-0636-7
1. Cats. 2. Cats—Behavior. I. Title.

SF447.W653 2007
636.8'0887—dc22 2007020703

2 4 6 8 10 9 7 5 3 1

Published by Sterling Publishing Co., Inc.
387 Park Avenue South, New York, NY 10016

Distributed in Canada by Sterling Publishing
c/o Canadian Manda Group, 165 Dufferin Street
Toronto, Ontario, Canada M6K 3H6
Distributed in the United Kingdom by GMC Distribution Services
Castle Place, 166 High Street, Lewes, East Sussex, England BN7 1XU
Distributed in Australia by Capricorn Link (Australia) Pty. Ltd.
P.O. Box 704, Windsor, NSW 2756, Australia

All photos from Shutterstock, Inc., unless otherwise noted.

Manufactured in the United States of America

Sterling ISBN-13: 978-1-4027-0636-3
ISBN-10: 1-4027-0636-7

For information about custom editions, special sales, premium and
corporate purchases, please contact Sterling Special Sales
Department at 800-805-5489 or specialsales@sterlingpublishing.com.

It has been said that “when the pupil is ready, the teacher arrives.”

Cats have taught me much about the essence of life and how to be true to myself and others. Integrity, charm, dignity, and style are just a few of their inimitable virtues, qualities which I still strive daily to incorporate into my professional and personal life.

I deeply treasure the lessons I received from my former cats: Muggsy, Sambo, Muggsy-Baggins, Sunny Blue, Honey Blue, Ziggy-Star-Dust, Diana-Moon-Dust, and my first cat, Oliver—master teachers of fulfillment all, who will live on forever in my work and heart.

Acknowledgments

Virginia Chipurnoi, Phyllis Levy, Barbara Myers, Susan Ferguson, Arthur Nelson, Francis I. Regardie, Frank Lerner, Harvey Gordon, Stark Beatty, my two sisters Emily Ouimette and Gail Mutrux, and to my friends, clients, and patients.

My special thanks go to Lee Fowler, my editor, for extending the life of this book, and to Claire Bazinet, who continued the process.

—Carole C. Wilbourn

A portion of the royalties from sales of this book will go to
The Humane Society of New York
306 East 59th Street
New York, NY 10022

Contents

Foreword

Carole Wilbourn understands that animals, like all living creatures, are happy or unhappy in direct response to the sensitivity and respect of those around them. Her book shows that, if their bodily needs are recognized but their emotional needs ignored, problems follow. For years, Carole has counseled at the Humane Society of New York, and her perceptive suggestions to animal lovers have preserved many rewarding relationships and prevented countless animals from losing potentially loving homes. Anyone involved with an animal can learn from her fund of informed, practical, and intuitive knowledge. We at the Humane Society of New York are also deeply grateful to her for turning over part of her proceeds from this book to continue on with our work.

—Virginia Chipurnoi

the Wilbourn way

I am repeatedly asked, "What *is* a cat therapist?" "What do you do?" and "How did you ever get into *that* line of work?"

I answer, as truthfully as I can, that I help people understand their cats' day-to-day needs, so that person and cat can relate in a way that promotes the emotional and physical health of both. Unfortunately, many of the cases I have seen over the years involve cats who developed deep emotional problems that precipitated behavioral disorders. Often the emotional problems also triggered medical problems, which is why I now feel that it is imperative, in my practice, to treat the "total cat."

Some years ago, Dr. Gale Cooper, psychiatrist and author of *Animal People*, explained: "The bedrock of the practice of psychiatry is the belief that people have inner lives, that their symptoms are manifestations of stress and that a therapist has access to understanding, by training and empathy. Problems are eased or cured by providing an atmosphere in which stress is lessened, so the patient can have the opportunity to learn new coping mechanisms. Sometimes medication is prescribed for relief of symptoms while the patient tries out new behaviors. Carole Wilbourn, a therapist, bases her practice on psychiatric principles. Her clients, however, are not humans, but cats."

I started my current "practice" by calling myself a behaviorist, but people thought of B. F. Skinner, and that connotation wasn't right. Although my methods are scientific and based on extensive observation (cause and effect), I follow the intuitive method in which I regard the cat as a *teacher* and *friend.* I impart the essential therapy to the cat's person to make the person aware of the needs and feelings of his cat—to get at the core of the problem—so the cat can be healed. In 1977, I found myself referred to as "the country's founding mother of cat psychoanalysis." Eventually, "cat therapist" seemed the logical and appropriate choice, descriptive of my work with cats.

It was in 1973, in Manhattan, that I co-founded the first veterinary hospital exclusively devoted to the care of cats. When Dr. Paul Rowan and I started The Cat Practice, I had no definite idea where my path would lead, other than that I had an overwhelming passion for cats and a desire to help and nurture them. During our time there, my observations and feelings about our patients became ever more important to our medical diagnoses. With my psychological and teaching background, I naturally began to speak to clients personally and to see patients with exclusively emotional problems on an individual appointment basis.

Since then, my "house calls" have led me to visit such places as Topeka, Kansas, Lake Charles, Louisiana, and even Kalamazoo, Michigan, to treat a cat named Bart and devise a program for his guardians on how to include their outdoor feral cats when they moved to a new home. I have also carried on phone consultations as far afield as Alaska, Australia, England, Italy, and Hong Kong.

Over the years, I have consulted with veterinarians to devise treatment programs for cats with psychological and emotional disorders, and from this has emerged The Wilbourn Way. Since 1985, I have been in residence at Westside Veterinary Center in Manhattan, where I see clients by appointment and work with hospital patients. My relationship with Westside continues to be both fulfilling and supportive.

I have served as consultant to The Humane Society of New York, on East Fifty-ninth Street, a privately endowed non-profit clinic and adoption center. The Society has added another significant dimension to my life. I wrote a monthly column for *Cat Fancy* magazine. For sixteen years, my "Cats on the Couch" column was a marvelous vehicle for dispensing cat behavioral advice and a great source of personal gratification.

Now it gives me great pleasure to be an online columnist for "In Defense of Animals" (www.idausa.org). IDA is an international animal protection organization dedicated to ending the exploitation and abuse of animals by raising the status of animals beyond that of mere property.

This brings me to my use of the term "guardian" in this book. I have long maintained that you cannot really own a cat. Yes, a cat will agree to live with you, but it will not be controlled by you. Cats will do what cats will do! So the term "owner," in connection with a cat, is out of the question for me. "Caregiver" or "cat's person" has been a better term. But, recognizing that cats have feelings, needs, and interests of their own, "guardian" seems an even more appropriate term for those who welcome such furry individuals into their homes, their families—and their hearts.

I hope to continue to take one day at a time, striving to provide the best of myself, The Wilbourn Way, for my patients and clients. As evident in this book, they deserve no less!

chapter one

do you really want a cat?

"I just don't understand! Tiger is nothing like what I thought a cat would be!" exclaimed Jean, my exasperated client. "I've had Tiger for just a few days, but he has me in a complete dither. He constantly jumps into my lap, my bed, and follows me everywhere!" She added, "Please tell me what I need to do."

This client's cat dilemma is not uncommon. Many people hold preconceived notions about cats that are just not realistic. I call these things they have read or heard "cat myths." There are many, many of these cat myths, and until someone has actually lived with a cat, that person often has many misconceptions about what cats are like. One of the most popular myths is that cats are aloof and independent. Tiger's person, Jean, had expected her cat to behave like a "mythical" cat; she was not in the least prepared for a clinging feline!

So let us consider: What is a cat?

As T. S. Eliot once wrote, "A cat is not a dog." Although both are domestic animals, the similarity ends there. A dog strives to please his master. A cat strives overtly to please himself first. This may make the cat appear selfish (which he is), but this selfishness is the key to his happiness. When the cat is pleased, he feels good; once he feels good, he may share his good feelings with his companions. (By the way, in this book it feels right to me to use the male gender when speaking of a cat or cats in general. This is not to be gender-partial, but merely to be consistent.)

Although a cat enjoys being praised, unlike a dog he doesn't hang on his person's every command or desire. A cat does not need constant approval from his person. This difference derives from the fact that a dog is basically a pack animal, thereby requiring a leader. A cat may have a companion, but otherwise it does not choose to exist in groups, and each cat is autonomous.

Life with a cat is filled with surprises and challenges. A cat is so very special that its person is forever learning new and creative ways to approach each day when sharing life with a cat.

What are some of the qualities that you feel make cats so unique?

Dignity, will, and contact. Cats will go out of their way to preserve their dignity. If you've ever laughed at your cat, I'm sure he gave you the cold shoulder until he felt you had redeemed yourself. The mere sight of a cat walking or preparing to sit down suggests dignity.

Cats are endowed with a tremendous will. Once their minds are set on doing something, you practically have to hold them down to change their course of action. A shout is not enough to divert the attention of a cat who's determined to do something. When it comes to life and death, a cat's will to live can withstand incredible odds.

Cats are exquisitely sensitive to contact, from both humans and animals. When I stroked my cat Sunny-Blue, his body would relax and fill with purrs. On many occasions the anxiety or discomfort I felt on a stressful day was dissipated by Sunny's comforting presence.

Cats are primarily controlled by their feelings and instinctively act to solve their problems in the most simplistic way. In other words, a cat acts the way he feels.

Are you trying to say that cats actually think all these things out?

No, not at all. The cat's senses are highly developed. These finely tuned senses enable him to identify and respond to many natural stimuli of which people are often unaware. The feelings in response to stimuli are more precise than reason.

Then you're saying that cats don't think?

No, cats have some cognitive ability, but they draw primarily on their feelings or emotions. It is this well-developed sensitivity that makes the cat the splendid creature he is.

Why do people say you can't own a cat?

Cats live primarily to please themselves. Although a cat may live with you and love you, he won't "belong" to you. His interest is in satisfying his own needs, and he seeks those things that please him. It's when he feels good that he also brings pleasure to his person, or those with whom he deigns to share his domain.

Are cats psychic?

Cats are natural mediums for fluctuations in surrounding energy fields—both human and animal. I like to call this ability "cat sense." If you already own a cat, you've probably had the following experience: You are by yourself in one room, with your cat apparently fast asleep in a room down the hall. Suddenly, you feel drowsy so you stretch out on the sofa. Almost simultaneously, your cat appears out of nowhere to stretch out next to you—or curl up on you! Attracted to your relaxed energy force, your cat simply decides to join you and share it.

Cats are usually attracted to mellow, moody, or even depressed people because they are nonthreatening. Repelled by violent or loud energy, a cat can feel your body expand when you're relaxed and your body contract when you're angry. That is why your cat is often drawn to you when you're happy, and even when you're sobbing, but will retreat or hide when you're angry or under stress.

Cats are so sensitive to energy field fluctuations that they can often sense an internal, organic change before it's evident externally. For example, a cat may suddenly, without provocation, attack or reject a close companion because he senses his companion is sick. He is able to feel his companion's discomfort and is overwhelmed, confused, and/or feels threatened. His companion becomes a source of anxiety, triggering the erratic behavior.

Mama Kehoe is a cat who had a very close companionship with her son

Chris. When he became critically sick, she continued to groom and stay as close to him as she had before. But toward the end of his illness, when he was outwardly restless and uncomfortable, Mama kept her distance.

Cats sense a similar fluctuation of energy in those who dislike or have a fear of cats. Cats can feel this anticat person's intense energy and may be drawn to its source. That is why it's not uncommon for a cat to plant himself in such a person's lap and knead away. The cat kneads to dispel the negative energy; the kneading makes him feel better. Of course, not every cat can handle such high energy. Some cats are driven away by it.

Someone who is anticat may also be allergic to cats, either physically or psychologically. If the psychological allergy is based on fear, it may be altered by positive will, if the person is placed in a position where it is necessary to tolerate a cat. If the allergy is an organic, physical one, it generally cannot be altered, at least not easily.

What's another cat characteristic?

Another significant characteristic is the cat's demand for special attention. Cats are such self-possessed creatures that they do not try to please us with their behavior but, instead, work at getting us to please them. In fact, they do it so well, we often feel honored by being able to oblige them.

Sunny-Blue, for example, adored special attention. His bowl could be filled with food, yet he would sit beside it and scream when I entered the kitchen. If I bent down and stroked him, he'd quiet down and nibble at his food. But at other times he would continue to cry until I finally figured out what he was craving.

Jean, Tiger's person, was completely overwhelmed by his demanding behavior. She had found Tiger crying outside her building. There was a note around his neck that said his person was forced to give him up. Tiger was two years old, neutered and affectionate. Jean decided to take him in because he was so pathetic; she didn't feel she'd have to give him much attention.

Jean's previous cat experience had been as a child. She had lived in the country, and her family had a cat that spent most of its time outdoors and rarely came home. Although Jean now lived in an apartment, she figured that most of the time Tiger would amuse himself. How little she knew!

Tiger was a real "people cat." "He can't seem to get enough of me," Jean cried. "I feel like I have this inescapable shadow in constant pursuit. Is Tiger's behavior normal for a cat?"

I explained to Jean that Tiger's behavior was actually very appropriate if you looked at it from his point of view. Unlike her childhood cat, who entertained himself outdoors, Tiger was used to interacting with people and was dependent on them for recognition and affection. Tiger was especially demanding of Jean's attention now because of his traumatic abandonment. As he became more secure and began to feel that Jean wasn't going to leave him, his need to be loved and noticed wouldn't be as dramatic. However, he would still depend upon her for attention, so I suggested a number of simple games she might enjoy playing with Tiger, such as dangling a string for him to bat at, tossing tinfoil or tissue-paper balls for him to chase, and giving him a paper bag or box to hide out in. I also recommended that Jean talk to him softly and repeat that he was with her to stay. Tiger wouldn't understand her words, but he would be soothed by her tone and the relaxed feeling of her body when she talked.

My next conversation with Jean was a pure delight. "Tiger has changed my life!" she exclaimed. She went on nonstop about how she couldn't wait to get home from the office to see him, how he had the most beautiful purr, and that she had never realized that there were so many cat people. Well, Tiger had wrapped his paws around Jean's heart, and the two of them would enjoy some happy times together.

Because adding a cat to your life can bring about so many changes and surprises, as in Jean's case, there are many considerations to be made before you make your "cat plunge."

A potential cat person should realize that adopting a cat is a total commitment. A cat's emotional and physical needs must be considered. For example, if you are a very neat person who must have absolute order all the time, you wouldn't welcome a cat and would welcome even less a litter box and loose cat hairs. A cat wouldn't satisfy a confirmed dog person, who wants the energy and attentive characteristics of a dog. An impulsive person who frequently changes his mind along with his lifestyle would be a questionable cat person, as would someone who is literally never home. Because a cat is a living being with feelings, the best type of potential cat person is someone who is willing to accept the responsibility—emotionally, physically, economically, and spiritually—along with the pleasure.

What's the age difference between an adolescent, a young cat, and an older cat?

A cat is an adolescent from about six to fifteen months. From fifteen months to two years, he's a young cat. After that, he's an adult cat. At about ten or eleven years, he becomes a senior cat. However, some cats with a high energy level and athletic body can behave like kittens up until they're eight years old. Lola, a ten-year-old tortoiseshell, darted about like a ten-month-old.

Is it a good idea for a working person who lives in a city apartment to have a cat?

Sure, as long as you can spend a few hours a day at home so your cat can receive an adequate amount of attention. However, check with your neighbors to see if cats are permitted in your building, if you haven't already noticed cats in windows or telltale traces of kitty litter by the incinerator. Two cats would be even better than one.

But wouldn't two cats require more care?

Yes, but you'd have double the amount of love and companionship that two cats would offer, and it's not twice as much work!

But doesn't a cat have to go outdoors?

A cat is definitely in his natural element outdoors, but he can certainly adapt to indoor city living, if you can provide him with adequate love and care.

I do a lot of entertaining at home. Will that present a problem?

Many cats enjoy visitors and will quite often position themselves to be the center of attention. If a cat doesn't care to socialize, he'll seek out a quiet out-of-the-way spot. You can always fix up a cozy nook in one of your closets for him to retreat to when guests are around.

I was thinking of adopting a cat, but my job requires me to travel a lot and I'm seldom at home. What do you suggest?

Perhaps you can take your cat along with you. If not, have someone stay with your cat while you're away. Otherwise, it might be better for you to visit your friends' cats when you're home rather than getting one of your own, if your time at home is so scarce. (See Chapter 11 for more information on travel.)

I already have a canary and a dog. What will happen if I introduce a kitten or cat?

Cats are natural hunters, and your canary would be a definite lure and temptation. Hang the canary's cage out of reach or on the wall in a place that a cat couldn't possibly reach by any means. If your canary is in the habit of playing outside his cage, don't adopt a cat.

A kitten can adapt quite well to living with a dog. However, if your dog is large, adopt a kitten that's at least three months old, so your dog doesn't accidentally injure or kill the kitten. If you decide to adopt a mature cat, the adjustment will definitely take longer than with a kitten, but the relationship can work out. (See Chapter 6 for more information on introducing a new cat or kitten.)

Will a cat or kitten get along with my children?

Yes, but if your children are under nine years old, it would be best to start with a kitten that's at least four months old. If the kitten is too small, it will be too vulnerable to a young child's overindulgence or mistakes. An adolescent or older cat is fine with children of any age. However, don't adopt a cat over four years old unless it has already lived with children or has a very mellow disposition.

What about finding the right kitten or cat for my personality?

Not every cat or kitten is suited for every person. Here are some pointers to help you select a compatible feline companion.

If you're not interested in combing and brushing your cat frequently or regularly engaging a groomer, select a short-haired cat. Adopt a long-haired cat if that's what appeals to you, but you will have to groom your cat on a regular basis. Colors have a definite effect on one's personality, so choose a cat whose coloring makes you feel good.

It's hard to generalize about sex differences and there seem to be even fewer differences once cats are neutered. So, unless you have a special preference for female or male cats, you can adopt either. Finally, are you excitable or laid back? Do you look forward to stroking a "lap" cat for hours, or prefer an active one who will entertain you with its antics? Test out your compatibility by visiting possible companions at your local shelter. Remember, if you adopt a cat that has reached sexual maturity (male between seven and twelve months; female between five and nine months or after first heat), you should have your cat neutered. (See Chapters 14 and 15.)

I'm single, live alone in a city apartment, and am gone most of the day. I'm not a compulsively neat person, and I don't mind some turmoil or mischief; in fact, I like high energy. I've never had a cat before—should I get a kitten or older cat?

Start off with two kittens; they should be at least eight weeks old. Two kittens can keep each other company while you're out and busy with other things. Also, if a kitten has a playmate, there's less chance he'll be lonely and run a destruction derby out of frustration while you're away. The sex of the kitten is entirely up to you. However, if you adopt kittens of the same sex, you can have them neutered at the same time.

If you'd prefer to have some help in your kitten's upbringing, you can adopt a young cat and a kitten so the kitten has a constant source of cat wisdom. Since you deal well with high energy, select the kittens or cats that appear most lively and active. The runt of the litter or a cat or kitten that appears shy and withdrawn would not be a terrific choice for your personality.

I live in a large apartment with three roommates. All of us like cats and have lived with them before. Although I will assume the most responsibility for the new cat, everyone has agreed to pitch in. One of us is usually at home during any time of the day, and my roommates are very particular about our furniture. In general, we've sought out quiet, well-mannered friends. What kind of cat would suit us best?

Since you've all had cats before and know the trials and tribulations of raising a kitten, I suggest that you select a young cat—at least fifteen months to two years old. Since there's still a possibility that a young cat may be high-powered and very mischievous, if you want to be sure of getting a more sedate cat, adopt a cat who is over two years old.

Even within these chronological guidelines, there are exceptions. Age is an important factor, but sometimes an older cat can be more vivacious than a younger one. It all depends on the particular cat's personality. So when you are choosing, see what kind of feeling you get from your potential companion. Ask the adoption center or the person from whom you are adopting the cat about its personality.

Again, it's usually preferable in a situation like yours to adopt two cats for built-in companionship. Kittens are most flexible in adapting to each other. If you're looking for older cats, you might try to adopt two who have lived together. It's usually more difficult to introduce two young or adult cats to each other, as they have more set patterns than kittens or adolescents and more easily feel threatened. But if you introduce older cats into your household at the same time, at least you won't have to deal with territorial feuds. Do make sure, however, that both cats are neutered to avoid some real cat fights!

Although you plan to accept most of the responsibility for your new friends, discuss with your roommates exactly how they plan to contribute in caring for the cats. This way you'll know more clearly what assistance you can expect. Your roommates may want to provide only affection and not any of the day-to-day cat care.

I'm single and live alone; I work at home and would like a furry, affectionate being to keep me company—a lap-cat who's gentle but somewhat aloof. I want to adopt only one cat, as I like to deal on a one-to-one basis. I've had experience living with cats before. What's your advice?

You'd do best with an adult cat between the ages of four and seven years old. Avoid kittens or younger cats unless you can find one that isn't too rambunctious and you decide that you won't be bothered by a young cat's natural growing-up characteristics. A younger cat generally has a lot of energy and might upset your routine more. Since you want only one cat, try to adopt a cat that has been in a one-cat family. Try to avoid adopting a cat who had a tight relationship with another cat and is used to cat interaction.

My husband and I have a large house, and now all our children are married and gone. I miss the presence of youngsters in our home, so I wonder if a cat would provide us with good company. I do enjoy taking care of plants and animals, and my husband is of somewhat the same nature. Since we've lived with dogs, and we've had only limited experience with other people's cats, a cat would be a new experience. What age cat should we adopt?

You're an ideal couple for kittens; eight weeks old is a good age at which to adopt them. Kittens are lots of fun and distracting. Also, since you like to nurse and nurture, a shy kitten or the runt of the litter might be just your cup of tea. Decide how much you feel you can offer in the way of rehabilitation. If you do decide on an introverted kitten, the second kitten should be outgoing, so some of his confidence will inspire the other. However, the second shouldn't be a super live wire or the shy kitten will be overwhelmed.

My husband and I live with our three children in a house in the suburbs. The children are one, six, and thirteen years old. Our family's never had animals before, but now we want to adopt a cat. What's the best cat for us?

I suggest you adopt a kitten between four and four and a half months old; choose one that's spunky and alert. You want to make sure the kitten's big enough to get away from your young children when it's had enough. Don't adopt a shy kitten, because the high energy level of your younger children might be too unsettling for him. Also, avoid the high-strung type whose stress tolerance might not be strong enough to cope with your youngsters; the cat might strike out in self-defense.

I might also suggest that you consider adopting an adolescent or young cat that's already lived amicably with children. And if your children are to take care of your new cat's daily needs, make sure they understand thoroughly what's expected of them.

I'm hot to adopt one cat or kitten. I work long hours, travel extensively, and when at home I entertain a lot. Also, I'm really a very neat housekeeper. Is my lifestyle compatible with having a cat?

It might be best for you to buy a toy cat, or cat-sit for one of your friends' cats when you're not on the road. However, if you insist you can't live without a cat, an adult neutered cat would probably be best for your lifestyle. Choose one that's lived in a one-cat family and won't miss interacting with another cat. A stray from the street might be perfect, because they generally are loners.

If possible, take your cat along when you travel. (See Chapter 11 on travel.) If you can't do this, make sure you have a cat-sitter or friend lined up who can fill in for you while you're away. Otherwise, your cat will get terribly lonely and possibly misbehave. Be sure to adopt a cat that's both social and mellow, so he won't be stressed when you entertain.

My family and I have rented a house for the summer, and our children want to adopt a cat or kitten. It's the perfect place for a cat! But should we adopt a cat under these circumstances?

Yes, but only if you're positive you can take the cat with you after the summer is over; don't be a summer-home cat family. What will happen to your cat after you leave? You can't assume that someone else will provide for him.

My husband and I are senior citizens. Our children want to present us with a cat, but we're not sure if it's the right thing for us. What do you think?

I'd recommend that you adopt an adult cat eight to ten years old or a senior cat eleven to twelve years old. A cat with a mellow personality should be a perfect lap cat for you. Avoid kittens because there will be too many growing-up responsibilities involved.

I have a two-year-old spayed female dog, and I want to adopt a cat, too. What kind of cat would best fit in?

Try to adopt a neutered cat that's had a positive experience with a dog, because a cat that has lived with a dog before will be more receptive to another dog.

If you want to adopt a kitten, adopt one at least fourteen weeks old and therefore strong enough to cope with the antics of a dog and less accident-prone. (Refer to Chapter 12, Introduction of a New Person, for more information on this subject.)

Will my place be wrecked by a cat? I take great pride in my furniture and rugs.

You'd do best with a cat that's at least five months old. A young kitten's activity might be too much for you and your furnishings. It might be best to consider an older cat, one that has a quiet disposition and has already been neutered. Then, buy a sturdy scratching post, so your cat can exercise his claws on his own furniture not yours. There are some good commercial posts out there.

If you prefer to create your own post, use a scratchy material, and line it with catnip. You'll find instructions for making one later on in this book. The important thing is to make sure the post is steady so it doesn't tip over when your cat scratches it. Place the post in an accessible corner. If your house is large, more than one post is advisable. To reinforce his habit, praise your cat elaborately whenever he uses the post, and start him with the post as soon as you get him. The quicker you acquaint your cat with a worthy post, the less apt he'll be to seek out your furnishings. Don't procrastinate and allow him to take matters into his own claws!

And what if he still scratches around?

Give him a spritz with your plant sprayer when he scratches on anything other than his post.

What about my ornaments and plants?

Move any fragile ornaments way out of reach, so they're not accidentally knocked over during a chase or by an ever-curious kitten who wants to make that top shelf.

Not all cats are plant eaters. Still, be sure to put your plants out of reach. If your new friend can't keep his mouth and paws off your plants, then hang them up.

Isn't there a substitute food for cats who eat plants?

Yes, buy some kitty grass so the cat has its own garden to nibble on. Unfortunately, not all cats enjoy kitty grass, but it's always worth a try, and it helps their digestion.

Is it best to start off with a kitten?

Not in every case, but a kitten would be good choice for someone who likes kids and has a high energy level. The kitten should be at least two months old to ensure his chances of being healthy and well adjusted. Actually, two kittens would be best so they could play and interact together.

Are there any disadvantages to starting off with a kitten?

Yes. Although a kitten can be delightful and entertaining, you must be prepared for spontaneous surprises. Don't rule out a sneak attack to your head and feet in the middle of the night or a kitten in your bowl of oatmeal. You may even be followed into the shower! Because a kitten will invariably scatter and rearrange your papers and accessible trinkets, you'll have to accept a certain level of disorder or not mind picking up and cleaning up.

What are the economic obligations connected with having a cat?

Economic obligations are greater with a kitten. A kitten requires several visits to the veterinarian for vaccinations, wormings, and neutering. After the first year an annual visit is usually all that's necessary for a healthy cat. Also, a kitten consumes more food during his first year because he's growing.

Does a kitten require more attention than an adolescent or older cat?

Indeed it does. You can expect a kitten to be all over you after you've left him alone for even a short period of time. Undoubtedly, he'll add some excitement to your sleep by either washing your ears or kneading your armpit or head.

What are the advantages of adopting an adolescent or older cat?

The first advantage is the built-in experience and savoir-faire that usually accompany age. Other advantages include only two feedings a day, less medical attention, and less chaos in your household.

Where can one acquire a cat?

Check out your local shelter or humane society. Those cats are usually caged and desperate for someone like you to take them home. Usually they are all good people cats. Pick out one that responds to you and to which you are attracted. You'll get a wonderful feeling knowing you've adopted an orphan, and your new cat will be off to a happy new start—thanks to you! The pet adoption column in your local newspaper and bulletin boards at your neighborhood supermarket are also excellent sources, as are friends whose cats have reproduced! You might also call your local veterinarians for leads. If

you want a special breed of cat, put an ad in the paper or your neighborhood grocery. The Internet is another source of animals for adoption.

How can I tell if a cat is healthy?

There's no way to be absolutely positive, because sometimes a disaster can be incubating and won't surface until later. However, make sure your choice is alert and has bright eyes and glossy fur. Watch to see if the cat scratches any part of its body insistently. That may be a sign of ear mites or fleas, problems that are easily remedied. Avoid cats with runny noses.

If you want an affectionate, demonstrative cat, don't choose one that avoids hand contact. However, a standoffish cat would be a good choice to introduce to another cat, because he would seek out another cat for attention before he would turn to people. Don't select a nippy or aggressive cat unless you have a companion cat or kitten for it to work out its high energy level.

You don't seem to recommend buying cats from pet shops or breeders. Why not?

First, I feel that since there are so many homeless and abandoned cats that need good homes, it's unnecessary to buy a cat. Also, many breeders and pet shops are in the business for purely economic gain and don't care much about each cat's individual welfare. There are exceptions, but usually the emphasis is on how much profit each cat will bring. It's simply not very economical to put a priority on the animal's feelings and welfare. The animal becomes a product—a cat for sale is as valuable to the seller as the price he's sold for.

For instance?

One of my clients, Ellen, bought a Siamese female from a breeder for ten dollars. While that sounds like a bargain, actually the breeder's generosity was based on La Put's inability to breed and her difficult personality. You see, La Put was already a year old, and the breeder had tried to breed her without any success. In addition to this, La Put was not people oriented, and the breeder's husband hated her.

What happened when La Put went to live with your client?

It was one dilemma after another for Ellen and La Put. When Ellen contacted me, La Put was urinating frequently on the bed and around the apartment, and was terrified of being held or petted. Occasionally La Put would accept contact, but only briefly, and then she would become frightened and hiss or strike out. I recommended that Ellen make an appointment to have La Put checked out and spayed.

How did she get her to the vet?

With the help of an anti-anxiety drug to make La Put manageable.

What happened after she was spayed?

Ellen weaned La Put off the drug, and she and her roommate were very giving with their love and care.

Did La Put get over her shyness and make any progress?

She did make headway, but unfortunately because of Ellen's situation the cards were stacked against her. La Put needed a very calm environment. Ellen's life was in a state of flux, as she was in the midst of applying to law school, changing her job, and moving to a new apartment. Because La Put's stress tolerance was inadequate, she couldn't handle all the extra stress from Ellen's unstable situation.

So what happened?

Ellen found a home for La Put with a retired couple in the country. They had the perfect situation for La Put. Ellen had given La Put her start, but they would be able to see her through.

Did Ellen get another cat?

After her life calmed down, she adopted Charcoal, a young female waif from the street. A few months later she adopted a kitten, and Charcoal's life became complete. (Go to Chapter 6 for more on this story.)

Selecting Your Cat's Name

Some people prefer to call their cat "Kitty" or "Cat." It's not unusual not to give a cat any name. Granted, a cat can get along perfectly fine without a name, but why should he? After all, he's special to himself and to you. Therefore, he should be referred to in a special way.

You may be able to name your new cat on the spot, or it may take you a while to decide on the right name. Perhaps you'll be influenced by his behavior or by the way he looks or even the way he moves.

Whatever name you decide on, it should make you feel good when you say it. Because when you say something that makes you feel good, your body relaxes and your voice and face reflect a positive feeling. The good feeling will be passed on to your cat and will make him feel good. If you choose a name that has an unpleasant or humiliating connotation, your body and voice reflect a negative feeling.

So when you give your cat that special handle, let it be one that rings with happiness.

chapter two

starting off to live happily with a cat

Is it correct to pick up a cat by the nape of the neck?

Kittens may be picked up by the "scruff" of the neck (the loose skin found on the back of the neck), but larger kittens and cats should not be picked up in this way, as it leaves all their weight dangling unsupported. You should always support a cat's weight from underneath when you pick him up, and you should hold a cat cradled in your arms. You can use his scruff to hold him steady, such as when he is having his ears cleaned, but be sure he is solidly supported from underneath.

What preparations and supplies are necessary for my cat's or kitten's homecoming?

First, close all your windows, so the cat won't make an unplanned exit. If you must keep them open, use securely fitting half-screens. Don't give your cat a chance to flirt with flight—they don't always land on their feet! As mentioned earlier, don't take a chance with your plants and fragile ornaments. Move them out of temptation's reach. You can offer some commercial kitty grass so your cat has his own supply of greens, or you can buy some seeds and grow your own. Plant two pots so that you can alternate them.

Be careful if you have an open fireplace. One of my clients adopted a young cat, brought it home, and let it loose in the living room. Suddenly her doorbell rang, and her new cat panicked and scrambled up the chimney. She telephoned me in great distress. I told her to check with her neighbor upstairs to see if her cat had appeared in his living room. Sure enough, her neighbor was quite startled to find a cat in his fireplace—and his two cats were quite annoyed to find an alien cat in their domain! This Houdini escape could have been avoided by starting the cat out in the bathroom, where he would have been more secure and sheltered, since it's a smaller space and less vulnerable to unexpected noises or people.

It's not unlikely that a new cat will jump or bolt at sudden or loud noises. Even if the home is not especially frightening, a new cat may hide out for several hours until he feels secure enough to deal with his new environment. One new kitten, named Monday, managed to crawl into a miniscule opening between built-in kitchen cabinets. The family worried that they might have to call the fire department to come to his rescue. After much coaxing, however, they were relieved when Monday finally crept out the way he crept in. The calmer you are, the more quickly your cat will adapt.

You'll need to buy a commercial carrier from the pet department of your local department store or animal-supply store. I recommend the model with the wire top. The wire top provides good ventilation, but the covering can be snapped down if your cat becomes frightened upon looking out. The carrier with the plastic top can become too hot. A company named Sherpa makes very human-friendly soft carriers that you can carry over your shoulder (long the accepted pet bags of airlines). A cardboard carrier is only a temporary conveyance, and some cats don't adapt to canvas or soft-sided bags because they are not solid enough. Line the carrier with strips of newspaper in case the cat gets nervous and an accident occurs. Left unlined, the carrier may become incurably soiled.

Then buy a plastic litter pan or dishpan and a plastic scooper. There is an electric self-cleaning litter box, but it's not every cat's favorite. It can unnerve the cat and cause litter box problems. There is also a covered box that cuts down on "litter spillage." Some cats prefer privacy, but others like to be confined. There are many litters to choose from. I prefer the clay and the pellet type litter. Some brands make litters that are low in dust. Organic litters made from recyclable ingredients are also available. Clumping litters can also be found, but sometimes the litter can adhere to a cat's rectum and cause problems.

If possible, keep the litter box near the toilet—if your plumbing allows, you can flush away some dirty litter. Scoop out the used litter frequently, and empty and wash the box at least once a week with a mild detergent. You may have to do this more frequently during the summer or if you have more than one cat. Provide an extra litter box for mutual convenience.

Provide one or two bowls for food and one for water. Avoid plastic bowls, because sometimes the hydrocarbons in the plastic will react with the skin to cause a rash under or on the cat's skin. Lead-based bowls can also irritate the skin.

Buy a premium commercial cat food that contains beef or beef by-products. Organic meats and poultry are good supplements, and dry food provides a tasty snack. Avoid tuna, because it is known to cause a vitamin E deficiency (in cats), skin problems, and increased nervous irritability. Other types of fish should be limited to an occasional meal. There are many desirable organic foods.

Your new cat will need a sturdy scratching post to exercise his claws and stretch out his body on. You can make a scratching post (see directions on pages 53–55) or buy one. Look for a post that is covered with sisal and filled with catnip. Log on to my website for more information. If you have a big apartment or multiple cats, get two or more scratching posts so your cats will find it convenient to use them. You can order or replace a post when it wears out (proof that your cat is using it).

What about toys?

Many cats like toys derived from choice household objects—anything from aluminum-foil balls to shopping bags to champagne corks. Most cats enjoy catnip toys and loose catnip. The choicest nip can be purchased from an herb store. You might prefer to grow your own catnip so your cat can enjoy a nip right from his own plant. (See sections at the end of Chapter 3 for toys and other things you can make for your cat.)

But isn't catnip harmful?

The occasional cat has an overly aggressive or nervous reaction to catnip, but otherwise it is simply a treat. Catnip is a wild-growing herb in the mint family whose active ingredient is not harmful. Most cats find it a pleasant aphrodisiac and a painless way to work off energy. Some "Victorian" or "Puritan-type" cats, however, are completely uninterested in the herb. (Actually, the love of catnip is an inherited one.) Should your cat turn out to be a catnip freak, you can keep his stash fresh in a sealed container in the refrigerator.

Is there any special comb or brush needed for grooming?

Buy a rubber brush for a short-haired cat. A wire brush and possibly a comb are necessary for long-haired cats who require daily grooming to avoid unpleasant matting of the hair. A grooming mitt is another option, or you can use a damp wash cloth. Loose hair will adhere to the cloth as you rub and stroke.

Is a collar necessary?

If your cat is going to be going outdoors, you should provide a collar and name tag. The collar should have some type of expandable safety release. An ID microchip is also available; it is implanted in the cat's neck under anes-

thesia. Your vet or nearby shelter will have more information. Kittywalk® Systems Inc. has an expandable netlike enclosure, which allows your cat to be outdoors safely. If your cat will be in contact with fleas, consult your vet about flea prevention; he'll need a flea collar or flea spray. A kitty harness is best if you'd like to take your cat for walks.

A kitty harness?

It's an adjustable nylon figure-eight harness that fits in front of and behind the shoulders. Cat strollers are also available, should you want company but your cat is not inclined to walk.

What kind of bed is best for my new cat?

A wicker basket is usually the most appealing, although many cats are happy with a comfortable-sized box or a soft chair. You may line the basket or box with either tissue paper or something woolen. Place the basket in a sunny spot, so your cat can lounge comfortably in the sun. It's not unusual for a cat to hang out in his basket during the day and share his person's bed at night. There are many variations; my sister Gail's cats, Spats and Raggs, slept the

night away in their matched wicker baskets on top of a bedside closet, but they preferred the bed for their morning snooze.

But isn't it unhealthy to let a cat sleep with you? I've heard that cats can take your breath away.

No, it isn't unhealthy unless your cat is covered with street dirt. He'd have to sleep right across your mouth to interfere with your breathing—that's another one of those fallacious cat myths. If that were the case, you'd move him to another spot. You can usually persuade your cat to sleep where you want it to on your bed, although some cats just love the pillow and others the foot!

Suppose he insists on sleeping on my head?

If it's your head he prefers, move him to another part of the bed and stroke him until you both relax. Donald Junior was a cat who insisted on sleeping every night across his person Susan's head until she provided him with his own special alternative. When she had a custom-built bed built into the wall, she instructed the carpenter to build a special shelf for Donald directly above her head. Although Donald enjoyed his own sleeping berth, Susan's head was not off-limits to him.

I've heard cats are nocturnal. Will my new friend keep me awake all night?

To avoid any midnight performances, offer your cat a snack or a pinch of catnip before bedtime. If his stomach is happy and the nip helps him work out his pent-up energy, there's a better chance you'll both sleep soundly. A brisk bedtime chase or "crazy hour" will also help the cat sleep better.

Is a medical checkup necessary when I get my new cat?

Yes, your adoptee should be examined by a veterinarian and given whatever vaccinations are needed at that time. His ears should be checked for mites and his coat for fleas. A stool sample will indicate if there are any parasites. It's best to have your adoptee's progress followed right from the start. (See also "Quick Cat-Health Pointers" at the end of this chapter.)

Because your cat is so greatly affected by your interactions, it is very important for your life together to get off to a good start.

How do you suggest I go about this?

Once you've acquired your new friend, talk to him softly and slowly on the way home. As you talk, think pleasant thoughts, relax, and breathe freely.

What do you mean? How can what I'm thinking about affect my cat?

A cat is sensitive to a person's tone of voice and body posture and naturally empathetic to feelings and fluctuations in energy. If your body and voice are relaxed, you pass on your positive energy. But if you're anxious, your cat can sense and feel your tension through your body language, which reflects it. Your cat becomes a perfect target for your anxiety.

So you're saying it's not easy to hide my feelings from my new cat?

Exactly!

Will a kitten be affected in the same way as a grown cat?

A kitten is also affected by people's behavior, but because a kitten is more flexible and his concentration span isn't as great, he can be distracted easily.

While carrying your kitten home, especially if he's high-spirited, his curiosity and attention may be centered on things around him and less on you. However, it's still necessary for you to radiate calm and happy feelings.

What do I do when I get him home?

Place his carrier in the bathroom next to the litter box, close the bathroom door, and then open the top of the carrier so your new friend can come out at his own speed. Introduce him to the litter box by gently lifting him into it and scratching his front paws in the litter.

How long must he stay captive in the bathroom?

Long enough to become familiar with it. After that, you can open the door and allow him to explore the rest of your place, one room at a time. This will keep him from becoming overly disoriented as he walks around smelling things out. Make sure he discovers his feeding bowls and scratching post.

What about food and water?

You can provide water but no food until he's comfortably acclimated. Don't take the chance of giving your new adoptee a nervous stomach upset.

Should I expect him to eat right away?

At first he may reject his food or hide out. If this behavior should continue beyond the first day or so, consult your veterinarian.

Would it be all right to have people come over?

A couple of people would be fine, but I wouldn't plan a gala party for your cat's homecoming. Allow him to become adjusted to his new environment with as little confusion as possible.

Will I be able to train him so he won't be destructive?

You can't really train a cat. If he feels like doing something, that's all the motivation the cat needs for doing it. The same applies for his not doing something.

Some cats, however, can be deterred by a sharp command of "No," a loud clap of the hands, or a spritz from a plant sprayer.

Then what are my choices?

You can prevent him from domestic mischief by anticipation. Don't leave piles of clothing or papers in his reach. Keep your closet door and drawers shut tight. Dishes in the sink or left on counters are a poor risk. Keep the garbage tightly covered and out of sight. If necessary, suspend your garbage can from the ceiling on a rope! Make sure you keep strong cleaning solutions, insect killers, and poisons out of your cat's reach. Any pins or needles should be hidden because cats are fascinated by shiny objects; if swallowed, these items can cause serious medical complications.

How can I stop my cat from doing things if I can't train him?

When he gets into something he shouldn't, distract him with a toy or a food treat that will tear him away from the forbidden.

Will I have to stay with him constantly the first day?

No, but when you do go out, say goodbye to your cat and reassure him that you'll be back. Be sure you see him "in-the-fur" as you say farewell to rule out the chance that he's gotten locked away in a closet or drawer where he'll be stuck during your absence. Be sure to leave a light on if it's dark or if you won't return until evening.

But aren't cats nocturnal? Don't they like the dark?

Yes, but they can't see in complete darkness.

What precautions should I take if a delivery or repair man has to enter my apartment while I'm away?

If the person has noisy work to do, shut your cat in an isolated room so he doesn't become terrified. In any event let the person know you have a cat and warn him to be careful not to let him out accidentally.

Must I close all my windows?

It's all right to keep them open if you have secure screens. You can open them a crack at the top or bottom if there is absolutely no way your cat can squeeze himself through.

I'm a night person and a late-morning sleeper. Suppose my cat insists on waking me up early for his breakfast?

If your cat's an early-morning screamer, cuddle him and speak to him softly and gently. Your low-key energy can help to calm him and pacify his stomach. If this doesn't work, get up and feed him and run back to bed.

What if my cat gets into my dinner as I prepare it?

If you've been out all day, greet your cat and tell him how beautiful he is. He won't understand the word beautiful, but he'll respond to the good, positive feeling you give off when you say it. Once you've checked his litter box, given him his share of cuddles, and given him his dinner, he won't be in the middle of yours.

What if he insists on getting in the midst of my cooking?

Try giving him just a taste. You're home free if he doesn't like it. If he insists on more, and he's eaten all his own food, give him more cat food. He's still hungry. You might also try to divert him with some catnip or a favorite toy.

And if he still persists?

Are you sure you've given him his share of attention? If he feels slighted, he'll do anything to get your attention. When I came home in a hurry and cut corners on giving my guys the attention they felt they deserved, they all but did a tap dance to let me know they felt deprived. Give your cat a good romp and play to avoid any attention-getting antics. But if you've exhausted all these solutions and he still persists, put him in another room. Usually, if he feels he has gotten his share of your attention and his stomach is full, he should let you prepare your own dinner and eat.

What if he insists on perching himself in the middle of my table?

Distract him by tossing one of his toys, clap your hands loudly and stamp your feet, or spray him with a plant sprayer. When my husband and I were having dinner, one of our cats frequently kept us company by stretching out on the table. If he got in the way, we moved him. Because we're vegetarians, he was seldom interested in our food. But occasionally there was something that pleased him, and we offered him a sample.

What if you have company for dinner?

Then we keep the cats off the table. Sylvester, a friend's cat, always plants himself in the middle of the table whether those seated are family or guests.

Isn't that embarrassing and annoying?

My friend tells me that they're so used to his table presence that they treat him as a "centerpiece" and hardly notice him.

Do rainstorms affect cats?

Yes, some cats are frightened by them. If your cat is, close the blinds, turn on some quiet music, and distract him with one of his favorite treats. Stroke him gently while you talk slowly and soothingly. If you repeat this approach whenever there's a storm, you'll have helped your cat to substitute a comfortable feeling for an unhappy one.

What about firecrackers?

These can be another source of anxiety. If so, follow the rainstorm approach. If your cat's still anxious, consult your vet about prescribing a tranquilizer (also referred to as anti-anxiety medication).

A tranquilizer? Really?

Yes, but be sure to find out from your vet what kind of reaction your cat might have to the tranquilizer. Initially, some tranquilizers can affect coordination, increase appetite, and produce disorientation.

Would you please be more specific?

If after taking a tranquilizer your cat loses his graceful presence and stumbles, talks nonstop, and eats furiously, relax and don't panic. Very soon your cat will flow with the effect of the tranquilizer, stop fighting the unusual sensation, and relax. (Refer to Chapter 8, Health, for more about tranquilizers.)

What about homeopathic remedies?

Yes, they are another option and have helped many cats.

Is there a way to relax my cat before a visit to the vet?

If your cat goes completely bananas at such times, you might ask the vet to recommend a tranquilizer. Give it a trial run before you schedule the appointment, to make sure it's effective. With or without a tranquilizer, the following instructions will help:

1. Keep the cat's stomach empty for at least a few hours before the appointment, to avoid a digestive upset.
2. Think calm and positive thoughts so you don't pass on any of your anxiety. Breathe deeply and exhale completely so your body stays relaxed.

3. Line the cat carrier with strips of newspaper to make it accident-proof. Sprinkle in some catnip to make it seductive. You might include a favorite toy.
4. If you have more than one cat, encourage the cats to play in the carrier so it contains their smell.
5. Gently cover your cat's eyes with your hand when it's time to put him in the carrier. Hold him securely but not rigidly. Stroke him and tell him that he'll be home soon. The associated good feeling will calm and soothe him. If the carrier opens on the end, you can hold it upright and lower your cat inside—hind legs and tail first.
6. When you arrive at the vet's, keep your carrier away from other patients so your cat doesn't become nervous. Talk to him and give him a few hugs if the waiting room is empty. You don't want him to be aware of other patients.

During your cat's checkup, keep your concentration on him and think good thoughts, so your positive energy is transferred to him. Keep your eyes off the needle if he's given an injection; you don't want to pass your reaction on to him. Remember, the more relaxed you are, the better your cat will feel. If you have to hold him while the vet examines him, use light contact, don't try to restrain him. Restraining him will cause him to try to get away.

The last time I took my male cat to the vet, my female completely rejected him when he came home. Why?

Because your cat picked up the smell of the other patients and the hospital. Your female was very threatened by these unfamiliar scents. To her, he smelled like an alien army of cats, rather than her own true companion. You can avoid such a situation in the future by rubbing the cat who stays at home all over with the lining of the cat carrier and allowing that cat to investigate the carrier. Then they'll both smell pretty much alike. If there's still a bad reaction, separate them for a while. I refer to this issue as post-vet angst.

My friend Phyllis once endured a major falling-out between her cats Barnaby and Tulip. When Phyllis returned home with Tulip from a visit to the vet, Barnaby totally rejected and hissed whenever she approached him. Tulip became sad and confused. She couldn't understand why her beau was giving her the brush-off. She now had a man problem, along with her gum problems.

Phyllis was beside herself when she called me for a quick solution. I told her to smear butter on Tulip's coat so that Barnaby would lick it off; this would break down his anxiety barrier. Phyllis replied that butter was not appealing to Barnaby, but he adored yeast flakes. Consequently, Phyllis sprinkled yeast flakes on Tulip's head at their next meal. Barnaby was hooked! A few more rounds and Phyllis reported that he and Tulip were back to their usual sex routine. Although both cats are neutered, each night they go through the motions, and evidently it's pleasurable for both.

On Tulip's next return trip from the vet, Phyllis sprinkled her with yeast flakes, so that when she greeted Barnaby, they had a loving encounter.

I can see there's a lot I'll have to do for this cat. Do you really think that living with a cat is worth it?

You're absolutely right. There is a lot you'll have to do to make your cat happy. But if you truly want to share your life with a cat, you'll have to do a lot of giving. At first you may think it's a one-way deal, but very quickly you'll find that your cat is giving you endless moments of love, entertainment, and companionship. As you interpret his needs and feelings, you'll find yourself becoming more aware of your own and other people's sensitivity.

Are you sure of this?

Absolutely! But if you are uncertain and feel you cannot devote your energy and attention to a full-time cat, why not volunteer some time at a local shelter? You may even have the opportunity to provide room and board periodically as a cat or kitten is awaiting adoption. You can be a foster guardian. A robotic cat that meows and purrs or just sits quietly on your lap for petting can also add some cheer to a catless home. (Consult my website for more information.)

Bedtime Jamboree

It's time to go to sleep, and your cat, who has been napping most of the evening, decides it's time for him to work off his energy, so he starts running wildly throughout the apartment. You shout at him to stop, but he continues to race across your body and under the bed. It wouldn't be so bad if this were an occasional performance, but it's the third night in a row. This one-cat show has got to close!

Here are some steps you can take to get your cat in the mood for bed and out of the mood for performing:

- Give him a late-evening snack to warm his tummy and take the edge off his activity.
- If your cat is too chubby for an extra snack, feed him a little less at dinner time so the remainder can be saved for later.
- If he has been asleep all evening, tire him out with a chase or his favorite game so that he uses up his frantic energy. Otherwise, he'll demonstrate this energy when you're snug in bed and he has you as a captive audience.
- Catnip is a good thing to give him if it appeals to him. It will make him active for a while and then he'll relax. Follow it with a few hugs to release his pent-up tension. If your cat still persists in performing, lead him to a separate room, and tell him to have fun, you'll see him later. Do this in a light-hearted way so the room becomes a comfort instead of a discomfort.

Remember, especially if your cat is young and has had a quiet day, there's a good chance he'll make up for it before the night is over. If you anticipate this reality, you can avoid a bedtime jamboree!

Early-Morning Food Call!

Last night was another late night, and you had every hope of catching some extra winks this morning. You even gave your cats an exceptionally ample bedtime snack so that their breakfast could be delayed.

But early this morning you made a very basic but blatant faux pas.

Nature called, and your trip to the bathroom was the cats' signal for breakfast. You tried to crawl back into bed and draw the covers up over your head, but this didn't protect you from their meows and tugs. You tried to put mind over matter so you wouldn't hear or feel them, but no good!

Finally you dragged yourself out to the kitchen and presented them with their breakfast. Now you were dismissed and could amble back to bed, but, alas, you were wide awake.

Here are some steps you can take to ensure your sleep:

- When your cats try to rouse you with their early-morning serenade, talk to them softly and try to lull them back to sleep.
- If that doesn't work, when they rub up against you so you can't ignore their presence, grab them and hold them tightly next to your body so they can't get away. Relax your body and stroke them so they forget their hunger pangs and drift off to sleep.
- Close your bedroom door, if all else fails, and tell them breakfast will be in their bowls soon. Cats are very habitual, so they basically expect their meals at the same hour every day. What you reinforce is what they will eventually accept. But their stomachs always want morning to come a little early.

If you find that none of the above helps and you want to sleep late the next morning, knowing that their inner alarm will go off before your radio alarm, prepare their breakfast before you go to bed. At the same time, be sure to give them a bedtime snack so they don't get confused and meow for their breakfast!

Quick Cat-Health Pointers

Vaccinations

A feline distemper vaccination is usually given to kittens between the ages of six and sixteen weeks. A kitten forms immunity against distemper somewhere during that time. Because it can't be determined exactly when the immunity forms, it is necessary to repeat the vaccination until the kitten is four months old.

The booster to continue the cat's immunity used to be given annually, but now, often, the booster is repeated only every three years. Feline distemper is an airborne virus—your cat does not have to come into contact with other cats or go outdoors to contact the virus. You can be the carrier!

Your cat should also be vaccinated against respiratory viruses. Consult your vet for the particular regimen and any other necessary vaccinations.

A rabies vaccination is necessary if your cat comes in contact with animals out of doors, or if you plan to take your cat to the country or abroad. A cat can contact rabies only if he is bitten by another rabid animal. In some communities, all cats are required to be vaccinated. If your cat comes into contact with other cats, a leukemia vaccination may be needed. Consult your vet for information.

Ears

If your cat's ears get dirty, you can swab them out (the part you can see—don't dig in!) with a cotton swab dipped in baby oil or lukewarm water. Wrap your cat up in a towel if he's a squirmer. Be sure to reward him after each cleaning so he forms a positive association. If in spite of your efforts the cat's ears become sore and remain full of debris, and you notice him scratching, he may have ear mites or an infection. Make an appointment with your vet to check him out.

Eyes

If sleep forms in the corners of your cat's eyes, you can clean them with a moist washcloth or a swab of cotton, always wiping toward the nose. However, if the eyes appear sore and runny, consult your vet.

Nose

You may also wipe your cat's nostrils with a damp washcloth or swab of cotton if they become caked or if his nose runs. But if the problem becomes severe, contact your vet.

Teeth

It's not unusual for a cat to have trouble with his teeth and gums—tartar built up on the teeth, gum erosions, bleeding gums, or occasional abscesses. A dry food snack or cooked chicken backs and necks are a good source of exercise for teeth and gums. Have your cat's teeth examined at least once a year. There are various dry foods that contain special ingredients to keep your cat's teeth healthy. I know of some cats who will allow their guardians to brush their teeth with a special brush and toothpaste. Sometimes the vet can simply scrape the tartar off with his fingers or with the aid of a dental pick, if your cat is willing. Otherwise, anesthesia is necessary.

Nails

Trim your cat's nails with a nail cutter to protect him against hangnails and your furnishings against sharp claws. However, make sure you don't trim higher than the red line. That's the claw's vein and cutting it will cause bleeding.

Gently press on your cat's paws as you trim. Present your cat with a treat after the pedicure, so he forms a positive association. If you have provided him with a sturdy scratching post, he won't have to claw your rugs or furniture. (Refer to Chapter 4 for more information.)

Parasites and Fleas

If your cat's eyes are runny, his coat is dull, or he has a potbelly, he may have parasites. Have your cat's stool sample checked annually for internal parasites. If your cat goes outdoors or has recently acquired a new companion, his stool should be checked more frequently. A kitten's stool should be checked each time he's vaccinated. Parasite eggs may not show up in the first stool sample, so have your vet check at least twice if you suspect your cat has worms.

If your cat leaves piles on your floors and avoids his litter box, he may have worms—or he may only be objecting that you are not cleaning the litter box often enough.

Regularity

To help his fur balls pass through and to aid his digestion, feed your cat a couple of tablespoons of butter each week. Some cats prefer a commercial gel such as Laxatone; or you may try occasionally adding a sprinkling of bran flakes to his food.

If your cat has chronic constipation, consult your vet. (See also Chapter 16, Senior Cats.)

An occasional bout of diarrhea may be remedied with yogurt or cooked rice in the cat's food. Also, limit the diet to bland foods, such as poultry and baby food, until the diarrhea passes. It may be the result of a plant the cat nibbled, but if the diarrhea persists, consult your vet.

Vomiting

Occasional vomiting may be due to a hair ball or a gulped-down meal. If vomiting persists, contact your vet.

Anal Glands

These may be a source of irritation to your cat. If he starts to lick or chew furiously at the base of his tail, his anal glands may be impacted. Have your vet check and empty them out, if necessary, on your cat's annual vet visit.

Bladder

A well-balanced diet and your attentive care will help to keep your cat's bladder healthy. (See Chapter 8, Health, for more information.)

Grooming

A short-haired cat should be groomed with a rubber brush. A long-haired cat needs more attention and a wire brush or various sizes of metal combs. If you can't groom your long-haired cat frequently, you may engage a groomer to come to your home.

If your cat's fur coat is dull and his skin is flaky in spite of frequent brushing, he may need more fat in his diet. Canned beef cat food, cooked meat or beef, and chicken bologna are good sources of fat. Butter is a good supplement, and you can also add to his food a supplement that contains the omega fatty acids, brewer's yeast, and wheat germ. You might find a humidifier effective, especially if in the winter your cat snuggles up to the radiator.

Sexual Maturity

A male cat usually reaches sexual maturity when he is between six-and-a-half and eighteen months old. You can tell by his strong-smelling urine and/or aggressive behavior when it's time to have him neutered. A female cat reaches sexual maturity at five-and-a-half to fifteen months of age. She can have an ovariohysterectomy after her first heat. This surgery is now being done at an early age by shelters so that the adoptees can't reproduce and add to the homeless cat population.

Wanderers

If your cat goes outdoors or even likes just checking out your apartment building, provide him with a collar and identification tag. Don't assume that he'll never get lost.

Purr, an outdoor cat, once wandered over to a construction site and camped out for a few days before the workers discovered him. His companions, Suzie and George, appreciated a holiday from his type "A" catsonality, but his person was beside herself during his excursion. She was thankful that she'd provided Purr with an identification tag.

Sweeney Todd, however, was a restaurant cat at Café Loup who zipped out the door and was finally discovered by a doorman who saw a sign posted about Sweeney's "leave of absence." If he'd worn an ID tag, his people would have been less frantic. Stoley and Shadow were two outdoor cats that had an electronic cat door purchased from Hammacher Schlemmer in New York City. (It may also be available online.) It opened exclusively by a specially coded key worn on their collars, so that no other animal could pay a surprise visit.

Although Cappucino and Lola were two indoor cats who lived in a duplex high up on Telegraph Hill, they received frequent visits from a neighbor's cat, who perhaps wanted to take advantage of their view.

Tuck was a young cat who lived with his companion cat, Scooter, and people in the country. However, every weekend Tuck would disappear until late Sunday evening. One Saturday his people went to visit friends nearby who kept a weekend house. As they talked, Tuck appeared on the patio, and they yelled: "That's our cat Tuck!" Their friends countered with: "But Yeats is our cat!" They'd found him on their property, assumed he took care of himself during the week, and considered him their cat on weekends. Tuck's "bi-family life" continued until his weekend family closed their summer house and returned to the city for the winter.

chapter three

exercise and the great outdoors

"I'd love to adopt a cat," said Sally, a potential cat person. "I always lived with cats in the country, but I just moved to the city, and I think it would be cruel to keep one cooped up in an apartment. What are your feelings?" she added.

Many people share Sally's impression. True, cats that go outdoors have no problem getting their share of freedom, adventure, and exercise. They're in their natural state. But they are also vulnerable to natural hazards: cars, unfriendly dogs, other cats, and mischievous children, to name a few. Outdoor cats generally don't live as long as indoor cats, since an outdoor cat is subject to all kinds of accidents and incidents. Recreation and adventure are limited for indoor cats, but they're not targets for the perils outdoors. Your indoor cat needs regular exercise to be healthy and to live a long time. How he gets his exercise is largely up to you. An indoor cat often has to be quite creative to stay in shape and keep from being bored or frustrated.

A companion cat provides built-in exercise, because two cats will naturally romp together, chase each other, and wrestle.

What inspires activity for an indoor cat?

Cats will madly chase an imaginary object or ghost or even their own tails. This diversion frequently sent my cat Sunny-Blue on a furious chase. Many cats will bat a piece of aluminum foil, champagne cork, or pipe cleaner about for hours. If your cat indulges in these sports, make sure he doesn't try to swallow any of his equipment. A box or paper shopping bag is a favorite source of entertainment; jumping in and out provides excellent exercise. Some cats will chase shadows or rays of reflected light on the wall or floor. You can simulate the beam with a flashlight. A scratching post will allow your cat to give his body and claws a good stretch and workout. The floor-to-ceiling model is ideal for a long workout.

Will my cat need any help or encouragement to exercise?

Although many cats develop their own daily exercise pattern naturally, if you have a lazy and/or on-the-way-to-becoming-chubby cat, he may need your help. There are many simple games you can play with him. For example, you can throw toys for him to chase or retrieve.

I didn't realize that a cat would retrieve.

Yes, there are some cats who are fanatic retrievers. J.J. is one. J.J.'s favorite playtime is bedtime. When his people are settled comfortably in bed, he appears with his worn toy mouse. J.J. drops it right down between them on the pillow so there's no chance they'll miss it. If they try to ignore him, he fetches one toy after another and lines them up on the bed. One little mouse is easy to ignore, but a menagerie is enough to start one tossing. So J.J. happily fetches until they just won't toss anymore. By this time he feels he's had a fair enough round and curls up on the bed with his companion Felix.

Why do you think J.J. waits until bedtime to retrieve?

Because his people are relaxed, they're in a fixed position, and he has their undivided attention. He knows they're not going to budge, and all he has to do is persist.

Can a cat be intrigued by a television screen?

Yes, and I've known many such cats. Liza was a young cat who spent many a happy hour hanging out over the top of the television, where she dangled her head and paws down over the screen. She seemed fascinated with the movement and reflections that appeared on it. Sometimes she would bat away at her own reflection. She was, of course, also attracted by the heat of the television. Quite often her person Sean had to look around her to do his viewing.

Games to Play with Your Cat

If you have a one-cat—or even a two-cat—household, your cat will surely welcome a repertoire of cat capers, games in which you are included.

1. Leave your empty grocery bag or box on the floor so your cat can camp out in it. In his new hangout, dangle a string or one of his toys so he can spring after it. The old toys can take on a whole new dimension for your cat if he plays with them in a new spot. (See directions at the end of this chapter for making catnip toys for your cat.)

2. Give him your latest champagne or wine bottle cork. Toss it around for him if he's intrigued.
3. Some cats love to play "toss and fetch it" with an aluminum-foil ball, paper, or toy. It's usually a caper they learn when they're kittens. I feel that this game is related to their hunting instinct; if given the opportunity, they will bring their catch to their person. If you want to reinforce your cat's fetch-it habit, praise him after he fetches and follow up the praise with a toss. When you've had enough, distract him with some catnip or his scratching post, or, if all fails, leave the room.
4. Hang a toy from your cat's scratching post and dangle it back and forth so he'll jump up after it.
5. Your cat might like to play hide-and-seek. Initiate this by chasing him behind a piece of furniture. While he's under cover, run to another room, hide yourself, and call to him to come after you. Your up-energy will entice him. Remain quiet until he discovers you. Praise him and start to run after him, but stop, so he can make his getaway and then you can seek him out. If your cat's intrigued, this should keep you going for a while.
6. Many a cat likes to be tossed up in the air. However, if you decide to toss him, make sure you catch him perfectly.
7. If your cat has a large toy collection, keep it in a handy basket so he can go to it and have a ready pick.
8. If your cat likes to take strolls but doesn't have access to the great outdoors, accompany him for walks in the hall if he enjoys your companionship. Whether you go with him or you leave the door open so he can return at will, provide him with a collar and identification in case he gets lost. Many cats can't get enough of the Kitty Tease (resembles a fishing rod) or Cat Dancer, which you can purchase at any pet supply store or make yourself.

When you join your cat in capers, make sure you give off a happy and positive feeling. If you're blasé and distracted, he'll pick up the feeling, and it won't be any fun or challenge for him. It's hard to hide what you're feeling from your cat.

The Great Outdoors

Can I take my cat for outdoor strolls?

Yes, but be sure to put a kitty harness on him so he can't make the great escape. A kitty harness is a nylon figure eight that wraps around under the cat's shoulders. You can purchase it at a pet-supply store or order one (please refer to my website).

What's wrong with a collar and leash?

The harness wraps around under his forelegs and gives him support, while it gives you a better hold on him. A collar can uncomfortably choke him when he pulls on his leash.

Suppose he doesn't like the harness and won't walk with it on?

Start him off by walking indoors, and give him a treat when you put the harness on so he builds up a positive association. Be sure to repeat this several times so your cat's good feeling about his harness is reinforced.

How do I go about taking him out on a walk for the first time?

Wrap your cat in a towel or put him in his carrier so that you can carry him outside painlessly. Do this at a quiet time of day; his first strolls should be in a secluded spot so he's not terrified by the street noises.

Suppose he's still frightened after a few trips?

If that happens, forget the strolls and let him gaze out the window. Many indoor cats lose their inclination to go outdoors. This can be especially true of older cats.

I'm moving to the suburbs, and I want to give my cat the opportunity to go outdoors. He's been an apartment cat all of his three years. How can I introduce him to the great outdoors?

First, the cat should be neutered. Initially, you and your cat must get completely settled in your new home. Your cat should get used to the interior layout before you gradually start taking him for short outings, with or with-

out a kitty harness. Stay with him for several visits, until you are confident he is used to the surroundings outside your house. He should be familiar with how to retreat inside the house quickly if he must.

When can I let him go out unsupervised?

After several supervised jaunts. But start him off slowly—about an hour the first few times.

How do I get him to come home?

Feed him each evening just before sunset, so he'll return home before dark. For safety's sake and your own peace of mind, keep him inside all night.

What's the best way to let my cat go in and out freely?

Install a commercial cat door. But be sure to close it when he comes home for dinner so he can't go night prowling.

What about identification?

You can supply him with a safety collar with an elastic band in it, so if he gets caught on something, he can slip out of it. Either write his identification and address on the collar or attach a tag.

Suppose he keeps squirming out of a collar?

At one time, cats were tattooed for identification purposes, but now you can have a microchip put in under your cat's skin. Consult your veterinarian or locate a shelter for details. The homes of many lost cats are discovered thanks to this microchip.

I know there's an invisible fence available to mark my cat's boundaries but are there other options?

Yes, there's a net-like enclosure that can be expanded to the desired size so your cat can enjoy the outdoors safely. If you have a terrace, you can even set up a free-standing enclosure there. But someone should stay around to supervise. Refer to my website for more information on these and other Kittywalk Systems products.

Suppose my cat resorts to catching birds. How can I discourage that?

Attach a bell to his collar to announce his arrival to the birds. If you're lucky, your cat won't spring before his bell can ring. Discourage birds from congregating on your property by not feeding them.

Many outdoor cats cannot resist their natural urge to pursue a bird. Because cats have become such domesticated animals, we forget that they are born hunters. When given the opportunity, they will hunt whatever moves and catch whatever they can get.

Unfortunately, cats may not always devour their prey. This can make the hunting seem cruel, but the successful cat is often well-enough fed not to be hungry. Some hunting may be done for sport, but that, too, is part of a cat's nature. If something jiggles in front of him, he's tempted and gets carried away in the chase. He doesn't stop to remind himself that he's not hungry, and he will pursue the chase to the finish—unless he's distracted.

You may be absolutely done in with grief and anger if and when your cat presents you with his catch. However, he does it to win your praise and esteem and to show you how wonderful he is. He doesn't know that you're not going to eat it or even enjoy it. If you yell at him or punish him, it will only confuse him. Try to remember that, although it's not the present you dreamed of from Tiffany, the bird is your cat's gift to you.

Here are some pointers to help you deal with your cat's bird-snatching:

- Don't tempt your cat by putting bird feeders on your property.
- Don't put out food for rabbits if they're abundant in your area.
- Provide your cat with a bell and hope he can't spring before it rings.
- If the catch is still conscious, distract your cat away from it with his favorite treat; then return to help his victim.
- If his catch is already fatally wounded, distract your cat with a treat and dispose of his victim.
- If you get the feeling that your cat wants to eat his catch, pick it up in aluminum foil or newspaper. Carry it over to a secluded spot or outside, where you don't have to be a captive audience to his well-earned feast.

Whatever you do, keep breathing deeply and try not to scream at your cat. You'll only increase the hysteria level. Remember, your cat's an animal, and he's motivated primarily by his inborn instincts. It is possible for a cat to be raised with a bird or rabbit without treating it as prey, but this is because of domestic circumstances and is not a spontaneous reaction. Your cat can be taught to interact favorably with a household bird or rabbit; but outside, he's driven by his natural instincts.

Can he get sick from eating the bird?

There is a possibility that he can contract worms. Have your cat's stool checked every few months, so your vet can put him on worm medication if necessary.

What should I do if a bird flies into my house and my cat runs after it?

Distract your cat as quickly as possible, head him off to another room, and return to help the bird fly to safety.

This reminds me of an amusing bird story told to me by my friend Mary Daniels, author of *Morris the Cat*. One day, when she took her cats Ashley and Mama Bear for their annual checkups, a sparrow managed to fly in through the chimney. Her house was not quite catless, as she had recently rescued an ailing kitten, Smudgie, whom she was trying to rehabilitate. Mary's daughter, Roxanna, entered the living room to find the bird flying in circles above a worn-out, collapsed kitten. Roxanna managed to ease the bird out a window and console poor, tired Smudgie. It was lucky for the sparrow that Mary's grown-up resident cats were not on the scene!

Can a country cat who adores the outdoors adapt to an apartment?

Some can. It depends upon how much attention and exercise the cat gets to compensate for his loss of the great outdoors. If you feel such an adjustment would be too traumatic for the cat, it would be best to find a home for him in his favorite territory or in another country spot. But be sure his new people will give him the love and care he needs.

Can an ex–house cat make it on the street?

Although it's possible for an ex–house cat to adapt to street life, it's usually a complete jolt and source of stress to a cat's whole system. Because a cat is a creature of habit and prefers to exist within his own familiar environment, the trauma of the street can make him a sure target for physical and/or emotional disorders. That's why it's not uncommon for a street orphan who appears to be in fine condition to become sick suddenly after adoption. Such a cat was able to cope on his increased adrenaline flow on the street, but once he felt safe and secure, he relaxed, and his body gave in to its feelings of sickness and need.

Don't some cats do well on the street?

Yes, there are those cats whose "catsonality" thrives on street living, but most of those were born on the street and it is their way of existence. A true street cat is wary of people and keeps his distance. This instinctive behavior pays off, as it keeps him from potential danger.

How do you feel about street or feral cats being altered and then returned to the street?

I know of many people who trap street cats in humane traps, so they can have them neutered by the vet and, when they've recovered, return them to their street spot. When we had The Cat Practice, we took care of many such cats. However, if a cat showed signs of responding to human contact, we and the cat's sponsor rehabilitated the cat and subsequently found a home for it. A group called Neighborhood Cats has a "trap, neuter and release" program based out of Manhattan, and Feral Cats is an organization with a similar program (you'll find more such information at my website). In this process, the cat's ear is notched during anesthesia to indicate the cat is no longer "intact."

What's your recommendation for dealing with a stray cat?

First, check the cat to see if it has any form of identification, including a tattoo. If it has a tattoo, telephone your vet or local Humane Society. They can provide you with the telephone number of the tattooing organization so you can trace the cat's person. If there's no identification, post signs in conspicuous spots with a description of the cat and your telephone number. Also, notify your neighborhood shelters and animal hospitals.

Have your cat's vet examine your new friend to make sure he's in tip-top shape. If you're low on funds, maybe you can take up a collection from your neighbors. Check online or the telephone book for Pet Finders, an organization that helps match lost animals with their people. (You might also want to check Craig's list.)

If you have a cat at home, be sure to keep the orphan separate from your cat, and wash your hands each time you've touched the orphan, so as not to transmit any illness. Have the orphan camp out with a catless friend if it's impossible for you to bring him home. If days pass and you can't located the cat's person, put a "For Adoption" ad in your local newspapers and post signs at local stores. Make sure your copy is appealing and snappy, so your orphan sounds desirable.

Keep in mind that if you come across a stray cat, he may not want a home. There are cats that prefer the street. However, if you take a stray cat in, once you tame him, you are responsible for his welfare.

How do you feel about giving a cat to a shelter or rescue group.

Although shelters and rescue groups do their best to care for orphan cats, by their nature they're orphanages. It's hard for a cat to get enough individual care and love there. The level of emotional and physical stress is great, and cats can become ill and infected from other sick cats. True, many cats are adopted, but a shelter should not be your first option for an indigent cat. Try your luck first.

Things to Make for Your Cat

The really creative among us can come up with dozens of things to make for their cats, whether appreciated by them or not. But the following few projects can be considered niceties as well as necessities, and they are fairly easy to do.

Project 1: Homemade Catnip Toys

Many cats do not react to commercial catnip toys. Either the catnip is ineffective, the cat doesn't appreciate the toy's texture, or the toys are simply uninteresting. Here's your opportunity to personalize. If your cat is a catnip fan, you can supply him with his very own custom-made catnip toys. Simply follow these instructions:

Materials

- Cloth scraps
- Sturdy needle and thread
- Scissors
- Catnip and teaspoon
- Tissue paper or newspaper
- Plastic wrap

Instructions

1. Cut up scraps of leftover material into shapes such as fish, suns, and moons, about 5 inches long.
2. With needle and thread, sew around the shape just enough to leave an opening for the catnip and stuffing.
3. Carefully invert your shape with your fingers, so most of your stitching won't show.
4. Sew a happy expression on the face, if you're so inclined.
5. Pour or spoon one-third of a teaspoon (or a bit more) of catnip into the shape's opening.
6. Stuff your shape with tissue paper or newspaper (watch out for the print), so it will crinkle when your cat plays with it. You could add a few leftover scraps of material to fill it out if you'd like.
7. Stuff your shape to the brim so your cat has a healthy toy.
8. Sew up the remaining opening.
9. Sprinkle a bit of catnip over the toy, and wrap it in plastic wrap or tissue paper.

If you have a sewing machine, you can speed up the production and make

toys for months to come. You can also present them to your friends' cats. What's especially nice about these custom-made toys is that you and your cat can both take pleasure in them. They make you feel good because they are your creations. They make him feel good because he picks up your good feeling. But most of all, of course, your cat enjoys them because they're such terrific toys!

Project 2: A Special Scratching Post

You've had the worst luck with store-bought scratching posts. The last one was a real bummer. Your cat reached up to scratch and did a flip as the post toppled over onto him. There's no way he'll use that post again! You know he must have an adequate post, or he'll continue to sharpen his claws on your furnishings.

You can build your cat his own custom-made scratching post, one that's both sturdy and seductive, so he'll make his mark there instead of on your rugs and sofa. This scratching post consists of a four-by-four post mounted on a reinforced base.

Materials

- 4" x 4" x 29" pole or post
- ⅝"-thick 16" x 16" plywood base
- ⅝"-thick 6" x 6" plywood base reinforcer
- Sisal or carpeting from a fabric or carpet store. Buy several yards, enough to replace it when it wears out. Cats love to dig their claws into this kind of material.
- Several ounces of your cat's favorite catnip or fresh catnip herb
- Catnip toy and string
- Carpet tacks
- Wood glue
- Nails or screws
- ¾" x 3" dowel, optional
- ⅝" wedge, optional
- Hammer or screwdriver
- Drill (depending on hardness of wood)

Note: When buying wood, you might ask your local supplier to machine-cut the pieces to size (to ensure clean edges and square cuts). Screws are preferable to nails for this assembly, but require a higher level of experience and skill.

Instructions

1. Cover the post with sisal or carpet, attaching the material to the post with carpet tacks or any suitable nails. Before completion, drop some catnip behind the material for extra seduction.
2. Mark one end of the post, both sides of the 6-by-6 base reinforcer, and the upper side of the 16-by-16 base as follows: Draw lines from corner to corner to form an X. Each of these pieces is a square, so the center of the X will be the center of each piece.
3. Place the end of the post on top of the base reinforcer and center it, then draw around the post, marking where it will be fixed. Do the same on the underside of the base reinforcer.
4. On the underside of the base reinforcer, mark where four nail/screw holes will be placed. They should be halfway between the center (the center of the X) and the outer edge of the post. For nails, drilled holes should be slightly smaller than the nail; for screws, slightly larger.
5. Spread glue on the bottom of the post, place it in position on the base reinforcer, then turn the assembly upside down. Make sure that the pieces are still aligned, and drive in the nails—2-inch common nails are fine. Hammer them in lightly to start, then check the alignment to be sure before driving the nails fully home.
6. Spread some glue on the base reinforcer. Center the base on the assembly, then nail through the base into the reinforcer, using 1-inch nails.
7. Hang a toy from the top of the post.
8. Allow twenty-four hours for the glue to dry, and the scratching post is ready for use.

Note: Another, or additional, way of securing the post to the reinforced base is by dowel-and-wedge. Using a ¾-inch bit, drill a 2½-inch-deep hole into the

center of the post. Drill another hole ½-inch deep into the center of the reinforced base. Then coat a 3-inch-long dowel with wood glue and insert it into the hole in the post and the base. Hammer in a ⅜-inch wedge to provide extra stability.

If you want a larger post, simply adjust the measurements accordingly. You can even build a floor-to-ceiling model, if you are so inclined. Your cat will love it. Build additional scratching posts as spares (use up extra carpeting) or if you need added coverage, i.e., a large home or multiple cats.

Sprinkle fresh catnip on the post for extra attraction. Praise and stroke your cat whenever he digs into his post. Spray him with the plant sprayer when he digs into your furnishings. Replace the toy at the top of the post with something else when he tires of it. If he's reluctant to use the post because of his former unsatisfactory experiences with one, feed him beside his post and play games with him using the post as center of attraction.

Don't despair if you're not the carpenter type but haven't yet found an acceptable post in a local store; good ones are available. (Information available at my website.)

Project 3: Your Cat's Modesty Screen

If you're tired of kitty litter on your bathroom tiles and your cat prefers privacy but won't use a closed-in litter box, a modesty screen is your answer.

If you prefer to build a custom model, here's a construction plan for a simple three-cornered screen:

Materials

- Three ¼ to ⅜"-thick pieces of plywood, 10" by 16" or sized to fit litter box (allowing room for cat to turn around, as they do). Ask your wood supplier to machine-cut pieces to size, and also drill holes along the edges for the twine (see drawing).
- Eight lengths of sturdy twine or decorative cord
- Wrapping paper, sketches, or drawn designs to decorate screen
- Glue
- Drill and ⅜-inch bit

Instructions

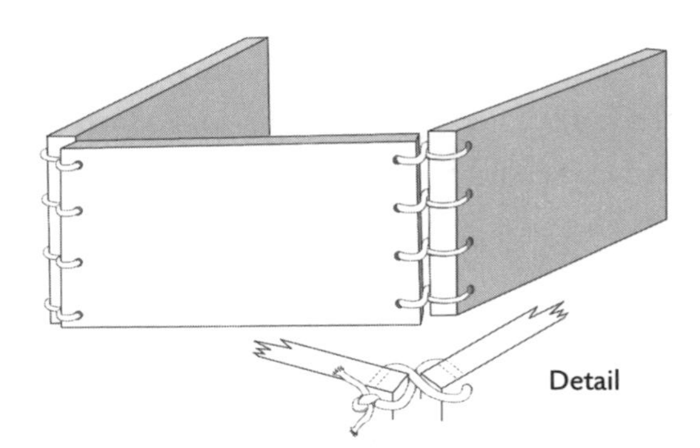

1. If your local wood supplier did not do it for you, drill the holes. The center board of the screen will need four holes along two opposite edges, whereas the other two boards will need four holes only along the edge that will be attached to the center board (see drawing).
2. After you've made your holes, glue on your decorative paper or get to work drawing your own scene.
3. Next, you're ready to loop the twine through each of the holes and knot. Voilà, your cat's modesty screen is a reality.

Place the screen around your cat's litter box, and stand back and admire your artistic work. Call or fetch your cat and introduce him to his new acquisition. He'll probably walk over and sniff it. He may walk away. If so, bring him back and place him in the box. When he's ready to use it, he'll either jump over one of the sides of the screen or push one of the sides away from the wall with his body or paw. At first he may be too modest to show you how much he's impressed by the new screen, but he'll indicate his appreciation by keeping more of his litter to himself.

How to Cope with Litter Box Problems

Fastidious toilet habits are the norm for a cat. Deviation from the norm is so bizarre and atypical that the cat's person, frustrated and confused, may conclude that the cat is avoiding the box out of defiance. That is not the case! A cat avoids the litter box to communicate physical or emotional discomfort. Unable to use words, the cat communicates through actions, which clearly speak louder than words.

Indiscriminate Defecation: Cause and Treatment

1. Untidy litter box: Scoop litter frequently and wash entire box at least once a week.
2. Objection to brand of litter: Return to former brand if cat rejects new one.
3. Dislike of type of box and/or its location: If your cat objects to a covered box, remove the cover. Keep the box in a quiet but accessible spot. A kitten can become confused and disoriented easily. Escort the kitten to the box, keep him company, and praise him when he makes a deposit.
4. Impacted anal glands: Excessive grooming in the anal area is frequently indicative of this problem. Make an appointment to have the anal glands checked and treated.
5. Constipation: This can sometimes cause a cat to leave segments of stool around. A laxative gel and/or a dab of butter will help soften the stool.
6. Onset of sexual maturity: An intact cat will often avoid the box upon sexual maturity. Untidy toilet habits will cease usually two weeks after surgery.
7. Retained ovary after ovariohysterectomy: An exploratory operation can be performed to locate and remove the ovarian tissue.
8. Parasites: Analysis of a stool sample and medication will relieve the problem. Coccidiosis is a common but difficult protozoan to isolate and identify. Giardhia is a very contagious virus that causes severe diarrhea but will respond to careful treatment.
9. Past injury to pelvis or adjoining area: Medication is usually the cure.
10. Cardiac or asthmatic condition: Medication and a support program are generally the regimen.
11. Anxiety triggered by an emotional and/or medical problem: A special support program and sometimes a tranquilizer used as auxiliary support are the answer.
12. Jealousy complex caused by a new animal, person, or baby in the household: Refer to my method of introduction and integration of new addition. My cat-caring CD may also be used for relaxation purposes (info on my website).

Indiscriminate Urination: Cause and Treatment

1. Sexual maturity: Two weeks after surgery "incidents" will usually cease.
2. Bladder infection or bladder stones: Medication will usually arrest a bladder problem, whereas surgery is the treatment for bladder stones.
3. Kidney disease: Involuntary leakage can be treated with medication and diet supplements. Sometimes acupuncture is successful.
4. Untidy litter box: Scoop litter frequently and wash out entire box at least once a week.
5. Need for privacy: Move the litter box to a quiet but attainable spot.
6. Visit from alien, intact cat (triggers territorial spraying): Contact cat's person to have intact cat neutered or keep it away.
7. Low stress tolerance—avoids box to communicate anxiety: A medical checkup to treat or rule out medical problems, initiate a support regimen, and, if necessary, prescribe a tranquilizer as auxiliary support.
8. Respiratory difficulties, e.g., heart, asthma: Medication and support regimen to reduce stress.
9. Jealousy complex because of new cat, dog, person, baby in household: Here again, refer to my method of introduction and integration of new addition. My cat-caring CD may also help bring some calm.
10. Discomfort from previous injury: X-ray and medication.

A cat's avoidance of the litter box is symptomatic of discomfort. Remember, don't consider the cat spiteful, but view his behavior with insight and apply a proper remedy so you can both live in harmony.

chapter four

claws vs. furniture

"I don't understand," said Joan impatiently. "Everyone knows a cat always lands on his feet." As Joan talked, she reluctantly shut her living-room window, and her cat, Marbles, jumped from the windowsill down to the floor with a very confused look.

Joan had asked me to pay a house call so I could get a look at Marbles and advise her on what type of kitten to adopt for him. Joan lived on the tenth floor of her apartment building, and she kept her windows open wide from the bottom; Marbles was introduced to me from his perch on the window ledge, overlooking Fifth Avenue.

I explained to Joan that although a cat was inclined to land on his feet, there were many exceptions. The variables were the decisive factors: If a cat landed before he was able to flip over, he would not land on his feet. Usually, if his fall was from above the fifth floor, he could not land safely. Even if he did land on his feet, he could be severely injured by the concussion of a steep fall. This calamity is known as the "high-rise syndrome."

"But Marbles has such wonderful balance," cried Joan. "He's too graceful to topple out the window," she added.

I agreed with Joan that Marbles was indeed graceful, but even so, his catlike grace was limited because he was declawed. The absence of his front claws inhibited his balance, no matter how he compensated for it. I told Joan that a cat's depth perception was poor, and all it took was one step too many and it could be curtains for Marbles.

"Maybe you're right," said Joan, "but Marbles has been hanging out on the window ledge for over a year, and he hasn't made any mistakes."

I assured Joan that Marbles had done very well but told her of a client's cat who had scampered around the fire escape for ten years until one day he made a fatal slip.

My advice to Joan was to purchase half-screens, making sure that they fit securely into the windows so Marbles couldn't push them aside (they could even be nailed). I suggested she fix up a comfortable seat for Marbles in her sunniest window, and if he showed any interest, she might even consider taking him outside for short strolls. A kitty porch, an enclosure that could be attached to the window, was another option.

Joan had another common misconception. Because Marbles was declawed, she was under the impression that his new companion would have to be declawed also, so as not to take advantage of Marbles.

I told Joan that she could put aside her fears. Cats usually keep their claws in when they're wrestling with a companion or friend. However, if the new kitten got too rough, Marbles would defend himself with his hind claws or by giving the kitten a playful but firm nip. His hind claws and teeth would compensate for his front claws in time of need.

So she'd feel even more confident, I suggested that Joan keep the kitten's nails trimmed to avoid accidents. I also recommended the she order a scratching post or floor-to-ceiling climber for the new kitten; Marbles would also enjoy it even though he was declawed.

"But my things are old, and I don't care if they get scratched," Joan protested.

I explained to Joan that although she didn't care now, if and when she purchased new things, then she would care. If her kitten formed good habits now, she wouldn't have to worry about changing its behavior later.

What do you have against declawing cats?

Cats are born with claws because they need them for defense, catching prey, balance, and grasp. Removing a cat's claws is unnatural, and the procedure is not without the potential for complications. There can be many physical problems. For instance, if the bandages are put on too tightly, gangrene can set in, and a foot may have to be amputated. Often hemorrhaging starts when the bandages are removed.

It sounds like these problems can occur with any type of bandaging.

True, but why subject your cat to an operation where he may suffer simply for furniture upkeep?

But once the surgery and postoperative period are over, he should be fine!

Not necessarily. Occasionally cats suffer long-term problems, such as abscessed nail beds or claws that begin to regrow misshapen.

How can this happen if the claws are removed?

Sometimes an entire nail bed is not removed, and some of the remaining claws begin to regrow. Also, it's possible for the bone to shatter, which can cause infection and continuous drainage from the toe. If this happens, the cat must be reoperated on, which is doubly traumatic to the cat, as well as expensive!

You mention the physical complications. Are there emotional problems?

Unfortunately the emotional problems are vast. The cat awakens from anesthesia to find his paws bandaged and clumsy. He feels confused and disoriented, and the pain in his paws causes depression.

Aren't you exaggerating the situation?

I wish I were, but I've observed too many victims, and although there were exceptions, most of the cats were quite miserable. Aside from the lingering pain, the cats couldn't stand up properly; they'd topple over their bandaged paws. Only after several agonized tries could they adjust to the bandages.

But weren't they okay after the bandages were removed?

No, then they discovered their paws weren't the same. When a declawed cat stretched out its paws, they felt stiff and they smelled different.

But the smells and stiffness soon disappeared?

Sure, but whenever the cat washed his paws, he couldn't stretch his claws in and out anymore, and therefore they didn't feel the same as they had before.

But these things sound like trifles. It doesn't seem like a cat's personality could be affected by such problems as these.

It's the anxiety that's triggered from these feelings that affects a declawed cat emotionally. He may become insecure and distrustful without having claws as his natural line of defense. So whenever he's in a new or threatening situation, he may overcompensate for his insecurity by becoming overaggressive. He becomes more apt to bite, since he can't scratch, and a bite wound can be a lot more severe than a scratch.

Also, a declawed cat's emotional reaction to the surgery can trigger various chronic physical ailments, such as cystitis, skin disorders, and asthma.

How?

The stress and anxiety that are the cat's reaction to the surgery can make the cat a vulnerable target for postsurgical problems.

What about something I've heard about called Soft Paws, that fit over a cat's nails?

The procedure requires anesthesia if the cat isn't cooperative, and most cats prefer unencumbered paws. Check out all the details before you make the decision.

Isn't there a modified declaw?

Yes, it's called a tendonectomy; the tendons that flex the digits are severed so the claws can't be extended, but they will still grow. This means that medical complications, such as dermatitis, can affect the nail bed. Also, why frustrate a cat with useless claws!

Okay, if it's unnatural to declaw a cat, isn't it also unnatural to alter a cat?

Once a cat has reached sexual maturity, the cat's physical and emotional health cannot be impaired by altering. In fact, if the cat is in an environment where he can't procreate, then altering removes the cause of his frustration and improves his health. (See Chapters 14 and 15, Sex and Breeding.) But a cat needs and uses his claws to cope on a daily basis.

I know of a cat that was hit by a car and had to have a leg amputated. He has suffered no aftereffects, and his personality isn't screwed up. Why would simply removing a cat's nails create the havoc that you mention?

Generally, before a cat has a leg, paw, or tail amputated, he has experienced discomfort, pain, and frustration with that particular part of his body. Once it is removed, the source of anxiety is gone, and he can quickly adapt to the new situation. But in the case of a declawed cat, his claws were never a source of pain or anxiety. Therefore, his association with them was not negative. When he awakes from the anesthesia, he experiences confusion, discomfort, and pain. Before the surgery, his paws felt perfectly fine.

This may all be very true, but I know of many declawed cats who are perfectly fine and have not experienced these problems.

So do I, but why take a chance by subjecting your cat to such an ordeal? Most of the declawed cats that I've encountered have an especially low stress tolerance and are overly tense.

But what can a person do if his cat insists on tearing his place apart?

He can make a sturdy scratching post. (See Chapter 3 for how to build a scratching post or where to order one.) It's best to have a few posts in a large house so there's always easy access to one.

How can I get my cat to use the post?

Compliment him whenever he uses the post, so you reinforce his habit. My late cat Sunny-Blue knew I would make a grand fuss over him whenever he scratched his post. One was stationed close to the apartment door. Whenever I entered, Sunny would run over and scratch his post, as if to say, "I'm such a wonderful cat, pet me!" Also, if he was feeling neglected or had created some mischief, he would head straight to his post.

And when he scratches my furniture?

Give him a spritz with a plant sprayer or water gun and yell "No!" very sharply and loudly. Don't be wishy-washy; mean it! Then take him over to his post and, when he scratches on it, praise and stroke him.

How else can I discourage him from using my furniture as a scratching object?

Cover it up until his post has become a strong habit and he forgets about your furnishings. Some kind of sticky tape can be found in stores. It will adhere to the sections of the furniture he scratches and the tape's sticky surface discourages claws.

My cat uses a special floor-to-ceiling post I built for him, but he still scratches my sofa. What should I do?

Cats love to climb up high. As for the sofa scratching, as I mentioned previously, use water to discourage him. Freshen up the post with catnip to entice him to use that. However, your cat may be scratching your things because he feels neglected and wants to get a reaction out of you. A cat who feels ignored will settle for any reaction, even a negative one. Try to give him his share of attention so he won't resort to mischief. Also, keep his nails trimmed.

If you feel you can't keep your cat unless he's declawed, find him another home where he won't have to part with his claws.

Pud is a senior cat who religiously used his custom-built floor-to-ceiling post. But when his person had a ceiling fan installed, Pud was terrified of the fan, and his post became off-limits. I told Pud's person to refer to the fan as a "Pud" fan, to feed him close to the fan and to start the fan off at a slow speed. Pud would acquire a positive association with the fan, and his fear would be defused. After several repetitions, Pud reclaimed his post.

Ian and Elsie are companion cats who work out on their post, but Ian is not partial and also seeks out the carpet. His people alleviated their "concern" with the purchase of a commercial carpet that is claw-proof but attractive.

Fire-Fly is a foxy Siamese who took a detour from her post to her person's upholstered chair. A spray of perfume to the chair, a new Felix post, and her detour was but a memory.

Duchess and Little J. are two declawed cats who use their post to give their bodies a super stretch and to exercise their back claws. Their younger companion, Geremiah, whose claws are "intact," was their role model.

Abigail and Louise are companion cats who frequently scratch away at their post in sync, chase each other about, and wind down with a snooze.

chapter five

your cat's emotions

I was having breakfast at my favorite Long Island weekend retreat when my waitress, June, mentioned that she had seen me on a television show. She asked me if a cat could have a nervous breakdown and went on to tell me how she and her husband had once had a seven-year-old Siamese cat with a wonderful personality. Everything was fine until they took in a female cat who had a litter of kittens. That's when the problems started; their Siamese withdrew, tried to bury himself in their bed, and gradually ate less and less. Finally she decided that if she let him go outdoors for the day, his appetite might be inspired. But the day she let him go outside was the last she saw of him. Although they searched and inquired, he had disappeared without a clue.

The Siamese had been their first cat, and at the time she didn't realize how much the mother cat and kittens disturbed him. She remarked to her husband that she thought their Siamese had experienced a nervous breakdown and that's why he left. However, her husband doubted that cats could have nervous breakdowns, and she didn't know enough about them to be able to figure it out.

I told her that her instinct was absolutely correct. Her Siamese couldn't deal with the energy level of the other cats or their competition for her and her husband's attention. He lost his will to remain with them and probably moved on to another place where he could get the attention he felt he needed. I added that a cat shows his feelings by the way he acts.

So many people are under the wrong impression that cats don't have feelings. It isn't until you've not only lived with a cat but taken the time to have a real relationship with your cat that you become aware of cats' emotions.

What do you mean, cats have emotions? That's anthropomorphizing—they're not people.

No, cats are not people, but they don't have to be people to have feelings or emotions. Cats can express happiness, sadness, rage, and anxiety, which are emotions.

Also, since they can't intellectualize their feelings, their feelings surface very quickly.

How can you tell which emotion a cat is displaying?

Once you become attuned to your cat's feelings, you will understand the varieties of body language your own particular cat uses to express his or her emotions. The easiest emotion to interpret is happiness, and the most obvious expression of it is purring, although purring doesn't always indicate that a cat's feeling happy.

Really?

Yes, when a cat's feelings become intense and his energy level is high, his purr mechanism may be stimulated, so that sometimes a cat can start to purr out of fear or anxiety. Jenny is a cat who invites you to pet her with a purr. Unfortunately, she soon becomes overstimulated, and her purr accompanies a swat and hiss. Her people really have to be aware of Jenny's mixed signals. (See "How to Recognize Your Cat's Emotions" at the end of this chapter.)

Like a person who cries because he's happy?

Exactly. Happiness can also be communicated by a cat's facial expression. There are cats who actually smile when they're happy. The muscles in their faces relax, and their eyes take on a happy glow.

I never realized a cat's face could be so expressive.

The better you know your cat, the more expressions you'll discover. Body posture also expresses happiness. When a cat is happy, his body is relaxed, his ears are cocked up and forward, and his breathing is calm. (Refer to Chapter 8, Health, for more on breathing.)

But a cat isn't necessarily happy if he's relaxed.

In order for a cat to be relaxed, he must be happy, but there are varying degrees of happiness and also different mixtures of feelings. For example, at dinnertime a cat feels happy, but he can also feel a touch of anxiety and anticipation as he awaits his food, whereas pure happiness might be revealed by a sleeping cat stretched out in the sunlight with his paws stretched out toward the ceiling. A cat who offers up his tummy as a vulnerable bull's-eye has to be feeling happy.

What are other ways to determine when a cat's happy?

Bumping or nuzzling—many a cat will bump his head or body against you when he's feeling happy. My late cat Baggins was a terrific bumper. If I held my hand up in the air, he'd spring up on his front paws and rub his head against my hand. I often measured his feeling of happiness by the height of

his jump. Other cats will nuzzle their heads against your hand or body to let you know when they feel good. Contact is another indication of happiness. Sometimes contact makes a cat happy; and other times, when he's happy he wants contact.

Do you mean contact with people or contact with other cats?

Both. When a cat rests his paw against his person's arm, this slight contact can fill his body with purrs. Or a cat's person can stroke the cat while he's already relaxed and purring away, and this contact will add to the cat's happiness. Also, a cat can provide happiness by licking his companion cat or increase his companion's happiness by nestling against him.

So you're saying that a cat can seek contact to make himself happy and also appreciate contact when he's already happy. Sometimes, though, when I pet my relaxed cat, he rolls over or moves away.

Well, there are times when a cat just doesn't want to be bothered, and he's entitled to his moments of privacy.

Don't some cats drool when they're happy?

Yes, but that is more of an individual trait; not all cats are droolers. Cats drool when their happiness stimulates their salivary glands, the way the sight or smell of food can stimulate a hungry person's stomach or chest, and still others will lick or suck their person's clothes or skin. This close contact is related to the intimacy and warmth they experienced with their mother during kittenhood.

I've heard of kittens and cats who insist on kneading people's chests and even sucking on their clothes. Why so?

I refer to these guys as "trotters and tasters." It's often because they were separated from their mother too soon and were not nursed long enough. Sometimes it's because they never got the pick of their mother's nipples.

Kneading stems from nursing; a kitten kneads his mother's breasts as he sucks on her nipple. This not only satisfies his hunger by stimulating milk release, it also gives him pleasure. The habit can be carried over to adulthood. When a cat or kitten comes in contact with a warm and comfortable person's body, the kneading action that once brought such satisfaction can be triggered.

But does the cat knead because he is happy or does the kneading make him happy?

Usually, it is because he gets a happy association, but it can happen either way—reciprocal.

How can you discourage a kitten or cat from kneading you?

If your cat's kneading really bothers you, move out from under. Or, accommodate him but concentrate on distracting him. My cat Baggins was a super "chest-trotter"; he would knead away until it felt right to him to sink down into the spot he kneaded. I would stroke and cuddle him to encourage him to relax and settle down. His kneading could be very piercing when his nails were due for their trim. Some "trotters" will substitute a cuddly sweater for a person's body.

And sucking?

You might see if a baby pacifier or bottle nipple will satisfy the craving. Sometimes it is indicative of a diet deficiency. Or it could be a nervous reaction to anxiety, which can be successfully treated with abundant attention and/or a synthetic or homeopathic tranquilizer.

What are some unusual ways that cats have of deriving happiness?

I could go on for pages with these anecdotes, but to name a few: hanging out on top of a doorway ledge, cavorting in the snow, riding in an elevator, munching on melon, eating pizza, being tossed in the air, snoozing in the baby's crib, and sleeping on a person's head.

You mentioned anxiety. Does a cat experience different kinds of anxiety?

Yes. There's fleeting anxiety, which a cat can quickly discharge; it is not threatening. However, prolonged separation anxiety that builds up and is not immediately discharged can turn into a pervasive fear or timidity which threatens a cat's peace of mind and physical well-being.

What's an example of fleeting anxiety?

Your cat's awaiting his dinner, you start to prepare it, and you're interrupted by the telephone. He becomes uneasy and worried that he won't get his dinner. His tail may flick, he'll meow and rub up against your leg, and his body may ripple to indicate he's anxious. But his anxiety quickly disappears as you hang up the telephone and return to his dinner.

How would you explain prolonged anxiety?

Prolonged anxiety can be caused by the presence of a new or alien person, cat, or other animal. For instance, a stranger may come to visit and remain a houseguest for a few days. Your cat may have an anxious reaction to the person's presence. Or your cat may have prolonged separation anxiety if he is left alone too much.

How can I tell if my cat is really anxious?

Your cat will communicate it by withdrawal, fasting, or even attacking the alien person.

You mean he might actually attack the person?

Yes, because the person would be the source of his anxiety, and he would attack out of self-defense. If your cat is having a prolonged anxiety attack, he's very prone to attack anything that moves, as it adds to his anxiety. Ankles are prime accessible targets of attack.

Are you trying to say that whenever I have a houseguest, there's a chance my cat might try to do him in?

No, that is the exception, not the rule. An attack is more likely to happen with a cat whose stress tolerance is low.

How can I calm my cat down if he does turn into an attack cat?

First, isolate him from your visitor and make sure that his isolation spot is peaceful, so he can calm down. If the anxiety does not go away, your cat may have to be tranquilized until he gets through his anxiety bout. You should also take pains to provide him with abundant care and attention, as the tranquilizer (anti-anxiety or psychotropic) would be the auxiliary support and your contact would be the primary source of help. (See Chapter 10, Catsonality Problems, for more information on difficult behavior.)

Do you find that aggressive tendencies are more apt to be present in some cats than in others?

Generally, a non-neutered, sexually mature cat is more apt to become anxious, because his energy level is higher and alien situations are more threatening to him. Also, a cat who has had a rough time on the street or who has been passed through many homes may have a difficult time dealing with strange people, cats, and other animals.

That's funny—it seems like a cat that has experienced many different things should be more adaptable and less threatened.

Some of these worldly types are. It all has to do with the particular cat's personality and how much stress he can comfortably integrate into his everyday life. Because cats are creatures of habit, they'll instinctively resist change unless they seek it out themselves.

Why is this?

A cat's familiarity with his daily patterns gives him security. Any major changes are a break from what he knows and can bring conflict and anxiety. Therefore, his reactions can be unpredictable and often unbearable. However, most situations can be resolved with time and patience from the cat's person.

Suppose my cat becomes anxious because of the houseguest's animal. What's the solution?

If the person and animal are going to be with you for only a few days, it's best to keep your cat separate from the visiting animal. It isn't worth the time and effort it might take to make their relationship civil. But if the animal is going to be with you for an extended period of time, the best thing you can do is to ignore the visiting animal, so your cat doesn't become jealous. (Refer to

Chapter 6, Choice and Introduction of a Companion.) However, don't invite a sexually mature intact cat to visit unless you have a young kitten or an extremely mellow cat.

You mentioned separation anxiety also. How might that affect a cat's behavior?

Your cat can experience separation anxiety because of the absence of a companion or person. If he is close to his companion or to you, his anxiety may affect his behavior. He may go off his food, hide, pout, or even get an upset stomach or a bout of cystitis. Willie, my friend Judy's cat, suffered a major bout of anxiety whenever Judy went away for summer weekends. He had his companion, Isabel, and several daily visits from a cat-loving neighbor, but Willie missed his Judy. One summer he slowly chewed off the fringe of her Oriental rug. Another time, he decided to lick away at his skin to reveal his anxiety.

If this could happen with even a short separation, what would be the outcome of a long or permanent one?

A long or permanent separation (such as the loss of a cat's companion or person) can deeply affect a cat's emotions and many times trigger a medical problem. The degree to which he is affected depends upon the particular cat's stress tolerance and his physical well-being.

Moon was a senior cat who lived with his people, Gail and Felix, a few cat companions, and Otis, an Old English bulldog. When Otis passed away from a critical illness, Moon followed shortly after. He was unable to endure the loss of his dear friend, and a chronic problem surfaced and decided his fate—to join Otis.

Can a cat ever be positively affected by a separation from his everyday environment?

Yes, it can occur. Spencer Tracy was a Siamese cat who tore away at his fur when he became anxious. He was forever on medication and tranquilizers. One time his person had to go away and left Spence and his companion with

Norma, one of our cat nurses. At first, Spence was disoriented in Norma's apartment, but he soon befriended her cats, who were used to cat visitors, and forgot about his licking fixation. He returned to his home with a healthy and happy fur coat. His experience in a new and neutral environment had dissipated his anxious tension and given him a "clean" furry start.

Do you feel a cat is affected by a marital split?

Yes; the stress can cause immediate or delayed emotional and/or physical problems. Sometimes a cat will reflect personality changes that are related to his person's new feelings of independence or freedom. It's usually best for the cat to remain with the person with whom he interacts best. If there are two cats, they should stay together, if their relationship is a close one. But if they must be separated, they will need a sufficient amount of comforting to ease the separation anxiety. After time has passed—several months at least—each would do well with a new companion.

It seems to me that a cat's person has to be really in tune with his cat to be aware of such problems.

Absolutely! But the more a person can open himself to his cat's feelings, the more perceptive he'll be.

What if the cat's person can't relieve the cat's anxiety in spite of being able to recognize the problem and spending a lot of time with the cat?

If a cat's prolonged anxiety can't be allayed by positive support from his people or companions, auxiliary support may be needed in the form of a tranquilizer (also referred to as an anti-anxiety drug). If the emotional anxiety has triggered a medical problem, they should be treated simultaneously.

How does a cat react after you give him a tranquilizer?

Reactions vary. Sometimes a cat's coordination is affected, and he may become wobbly within twenty minutes to an hour after the drug is given. He may feel disoriented, start talking, and try to resist the effects of the drug by running around.

What should the person do when the cat starts staggering around?

Relax, have a delicious snack, and try to have good feelings. If the cat's person goes bananas, this only adds to the cat's possible disorientation. Never laugh at your cat's reactions to a medication, even if he appears funny or clumsy. Talk softly and gently to your cat, and stroke him if he's receptive. Your cat's appetite may increase. If so, you can give him more food, within reason. After your cat's system adapts to the drug, his bizarre reactions are generally minimized.

How can I tell if my cat's dosage should be increased?

If he still appears overanxious—has a rigid body, cries constantly, and resumes the behavior you're trying to change.

But suppose the cat becomes addicted?

That is quite unlikely to happen. As the cat's stress tolerance increases to the point where he can interact on a comfortably sustained day-to-day basis, the tranquilizer can be decreased.

What happens if the cat has a setback?

Then the drugs must be increased until the cat can cope comfortably, then subsequently slowly decreased to the maintenance dosage.

But when should the drugs be eliminated?

The drugs can be stopped once the cat is integrated enough to interact and function without incident.

Sounds like it could go on indefinitely.

Well, the time element varies. It depends upon the cat's degree of anxiety, his person's support, and the cat's individual healing ability.

How does a person know when his cat can stop taking the drugs?

Usually a person can tell when his cat's stress tolerance has increased so that he no longer needs drugs. It's when his cat's maintenance dosage is down to a very small dosage and the cat is still snoozing frequently.

So you really don't think that a cat can be hurt by drugs?

They can't hurt a cat if they are prescribed carefully by a veterinarian and given sensibly. The amount of medication needed differs with each cat. It depends upon his particular stress tolerance and how well his body absorbs the tranquilizer. If a cat's stress tolerance is low and his body doesn't absorb the tranquilizer easily, he'll need a large dosage. There's no set formula as to how much any one will need to make him feel comfortable. Generally, a cat's blood should be evaluated to rule out a defective liver or other problem that could be escalated by such a drug.

Are you suggesting that a cat's uncomfortable when he's anxious?

If it's fleeting anxiety, there's no problem. But when a cat suffers from continued or prolonged anxiety, he feels uncomfortable and may even hurt inside. If the discomfort continues, it can often precipitate physical problems—the bladder, fur, skin, etc., could be affected. The particular stress target varies with each cat.

What is a stress target?

A stress target is that part of any particular body that is most vulnerable to stress and anxiety. When a cat has an anxious or rough time, the most vulnerable parts of his body are affected—similar to psychosomatic illnesses in people, which are real illnesses caused by emotional or mental upsets. (See Chapter 10, Catsonality Problems.)

You seem to have a lot of faith in anti-anxiety drugs for cats.

I do, if they are used prudently. If they are used effectively to relieve anxiety, the prognosis is encouraging and optimistic.

What if a client is skeptical about using such a drug?

I explain to the client that if he has doubts and fears about tranquilizing his cat, it would be best to find another way, because his uncertainty would interfere with the treatment. Homeopathic remedies are available as an alternative measure.

But is there another way?

Aside from removing the cat from the source of anxiety, it would take an incredible amount of time and patience from the cat's person and the chance of many setbacks. It would definitely be a long and trying route.

Would a maintenance dosage have to be increased at any time?

Absolutely. Extra stress, such as the prolonged absence of the cat's person, continuous noise, a houseguest, or an illness in the immediate family might affect him and cause him to have a setback. During any of these or similar stress-related circumstances, the maintenance dosage should be increased for a while and then reduced after the tension's over.

When, exactly, should the dosage be reduced again?

A few to several days after the status quo is reestablished, because it takes a while for the effects of the stress to diminish.

You mentioned that the anxious cat can sometimes attack the object or person that triggers his anxiety. Once my cat got mad at his brother but scratched me when I tried to separate them. What did I do wrong?

That was your first mistake. You should have let them box it out. If it had got too hysterical, you could have separated them with water or by throwing a blanket or coat over one or both of them. Unfortunately, you were a victim of your cat's displaced anxiety, so you got the brunt of his aggression instead of his brother.

Are there other situations in which a cat displaces his anxiety?

It's not uncommon for a cat to get mad or annoyed at one of his companions for what appears to be no obvious reason, and to take out his aggression out on a person.

Why would he do this?

If his anxiety wasn't fleeting but prolonged, and the source of his anxiety was a favored companion, a cat might transfer that anxiety to a companion or person to which he was less close.

My sister had a weird experience with her cat. The cat fell off the sofa, hissed a few times, and then scratched my sister's friend. Why did she do this?

The cat was upset and humiliated because of the clumsy fall; and transferred her anxiety and displeasure to the visitor.

Do you know of any similar cases?

Unfortunately, too many. One I'm reminded of is the bizarre reaction of a cat named Paulette. Her person's boyfriend came to visit, and Paulette's paw was caught in the door. Although not seriously injured, Paulette had good reason to be anxious and upset. However, she transferred her anxiety to the boyfriend. She ran and hid whenever he came over, refused to eat, and to top it off she urinated and defecated outside her box.

What caused her to react so dramatically?

Paulette was the only cat, which made her a victim of the "single-cat syndrome." (See Chapter 6.) Paulette had been declawed, which intensified her insecurity. And the boyfriend threatened and interfered with her relationship with her person. Paulette was so traumatized that she lost control over her bladder and rectum.

I see. So the boyfriend became her source of anxiety, and her stress targets were her bladder and rectum.

Yes, the boyfriend was a victim of Paulette's jealousy syndrome.

Did Paulette get over this displaced hostility and jealousy?

Yes. I explained to her person that it was vital to Paulette's well-being that she give Paulette more attention and have her boyfriend feed Paulette whenever he visited, so the cat could form a positive association with him. Also, they should try to include Paulette in their conversations so she wouldn't feel neglected. I also recommended that Paulette be started on an anti-anxiety drug to relieve her prolonged anxiety. Positive support would not be enough.

Ah, I see, the drug was the auxiliary support. What was the outcome?

Paulette's stress tolerance increased to where she could cope with the boyfriend, and her jealousy complex diminished. The anti-anxiety drug was slowly tapered off and then eliminated.

Will Paulette ever need anti-anxiety drugs again?

Possibly. How she'll react to similar situations depends upon how strong her stress tolerance is.

Do you think that a cat can be affected by his person's anxiety?

Indeed so. I often find that when I'm in an anxious state, my cats are an accurate barometer of my anxiety. For example, one day I arrived home with just enough time to give my cats dinner, telephone a few of my cases, and write a few pages before I had to dash off to class. My frenzy was so obvious that Baggins, who usually vacuumed up his food, only took a few nibbles, and Sunny-Blue ignored his bowl. Then the two of them got into an unfriendly squabble.

Well, my priorities shifted! I knew that if I tried to make any calls, Sunny-Blue would scream in protest and Baggins would claw away at the door. So I gave them each a few pats, took them for a stroll in the hall, and postponed my work until after class.

What do you think would have happened if you hadn't given them the attention they asked for?

If their anxiety had grown because they felt neglected, Sunny-Blue would have opened all the closet doors, batted the rubber duck into the tub, wrestled with the toy cat, and stalked Baggins.

What about Baggins?

Although demonstrative, he was less so than Sunny. Chances are he would have finished off his food to relieve his anxious feeling and topped it off with a snooze in his living-room basket.

So this would be an example of fleeting anxiety?

Yes; however, if the situation were repeated frequently, their reactions would become more drastic.

Such as?

Since Baggins was more reflective and introverted, he tended to internalize his emotions; therefore, he would probably develop a physical problem. His stress target was usually his anal glands, so they would probably become impacted and have to be emptied out.

And Sunny-Blue?

Though he was extremely verbal and extroverted, his anxiety would probably have precipitated a slight setback. Being a onetime street cat, he could revert to a modified attack cat and give a few light nips and hisses to communicate his displeasure and neglect. Also, since his body would contract because of his anxiety, he could also suffer a physical problem.

I try to be in touch with my cats' needs, but there are times when, like everyone else, I have to make a real effort to consider their feelings when I'm under various work or schedule pressures.

How does anxiety differ from rage?

When a cat is in a rage, there's no way to mistake his feelings. Generally, his ears flatten, his pupils dilate, his back ripples, his fur stands on end, and his tail jerks. These characteristics may be accompanied by an eerie, continuous yowl.

So a cat's rage is more dramatic and pronounced.

Yes, but it usually erupts quickly and doesn't tend to linger on. The rage disappears once the energy is let out, whereas anxiety, being a state of apprehension, fear, and/or worry, can linger a long time, sometimes indefinitely.

Do you have a good rage story?

Sunny-Blue, although daring and fearless, was quite intimidated by dogs. One afternoon while running his relay races in the hall, he almost collided with Candy, a neighbor's dog, as she came around the bend to her apartment. Then Sunny almost ran smack into Samantha, another dog, as she and her person were en route to the elevator. Well, this saturated Sunny's low dog tolerance. There are only two dogs on our floor, but they were two too many for Sunny. His back went up, his tail plumed out, and he practically flew into the apartment and gave a banshee yell as he tackled Baggins, who was just about to venture out into the hall.

Then what happened?

They wrestled and chomped at each other until Baggins planted his whole body on Sunny (Baggins is twice as big), and Sunny's rage was expelled with the pressure of Baggins's body. Baggins's weight caused Sunny's body to expel his pent-up breath, and as he exhaled, his fury dissipated.

What would you have done if their spat had continued?

Sprayed them with the plant mister and/or distracted them with a snack or a favorite toy.

You wouldn't have actively separated them?

No, when a cat's in the midst of a feud, a person should never get in the middle and risk the chance of becoming a victim of displaced aggression. The best procedure is to remove the source of rage, keep a safe distance, and distract the cat with something that will please him—replace his rage with pleasure.

How does a cat communicate sadness?

His eyes may take on a sad expression; his spirits may decline; or he'll generally deviate from his ordinary behavior. If it's a physical problem that has made him uncomfortable and sad, he'll usually try to draw attention to the source of the discomfort.

Do you really think a cat is sad when he's in discomfort?

Yes, although the sadness is mixed with discomfort and anxiety; he's unhappy because he doesn't feel good.

Does a cat ever reject his litter box out of sadness?

Yes. Again, discomfort would have triggered the sadness. Discomfort could be caused by diarrhea, constipation, parasites, or a bladder problem.

So when a cat is internally uncomfortable, his ordinary daily habits become weird?

Yes, they become abnormal or bizarre. That is how the cat communicates his discomfort. The problem can sometimes be a dormant one which is triggered by a stressful situation.

What is a dormant problem?

It's a low-grade, underlying problem that normally never appears but flares up and surfaces during stress. Cystitis is frequently a dormant problem with cats, but sometimes the condition escalates in frequency and seriousness to become a chronic problem. When a problem is chronic, the stress from the source of anxiety affects the most sensitive or vulnerable part of a cat's body that has been troubled before.

Can you give an example of a chronic problem?

Sammy was an adult cat who had suffered but recovered from a severe bout of pancreatitis. It occurred when Milton, one of his people, was afflicted with cancer. For many months Milton rallied, and all was well at home. But eventually Milton became worse; he was hospitalized with a terminal condition. Although his wife, Ruth, did her utmost to give Sammy and his companion, Charlie, extra love and support, there was a limit to her physical and emotional resources.

Sammy, inevitably, had a relapse of pancreatitis, was hospitalized, and for a while it was touch and go.

Did he pull through?

Yes, with the help of intravenous fluids, conscientious medical attention, and the strong contact from his person. But I think it was Ruth's bra that really helped Sammy's condition, because it added a touch of cheer and humor. She was asked to bring in something of hers that Sammy could smell to make him feel good. Ruth decided that nothing could be more intimate and personal than her bra!

So even though a cat's medical problem is treated, the sensitivity can remain and the problem can return for an encore?

Yes, a cat's stress target (the most vulnerable part of his body) is threatened in various anxious situations.

Does stress affect each cat in the same way?

No, it all depends upon each cat's stress tolerance and his general health.

You mean because a cat's emotional health can affect his physical health, just as can happen to a person? How can you protect your cat against such situations?

There isn't much that can be done in a case like Sammy's. Usually, however, if you can anticipate and minimize stress, the rate of recurrence will be lower.

I must repeat that a cat is a creature of habit and can easily become disturbed when his routine is altered. For instance, if you travel with your cat, a tranquilizer might be effective if he tends to be anxious. If you leave your cat at home, arrange for someone to come in to feed and visit at least twice a day. Cats thrive on human contact. Even if your cat ignores the visitor, at least it's positive attention.

Your cat is prone to depression when you are away. When his resistance is lowered, he's a vulnerable target for sickness. That's why it's so important for you to have someone shower your cat with special care during your absence. Increase your cat's maintenance dosage if he's on any kind of medication, whether he stays at home or travels with you.

What about a recurring problem for which he has received medicine in the past?

It may be wise to repeat the medication a few days before the trip and a few days after. Consult your vet for the dosage.

I suppose a houseguest might upset my cat's routine.

Your cat might feel left out. Try to give him his share of attention. It might be desirable to increase any daily medication.

Are there other reasons, besides medical, for a cat to feel sad and to deviate from his ordinary habits?

Yes, loss of or separation from a companion or person might cause him to express grief by constant sleeping, which he will do to get away from his sadness. If he's an outdoor cat, he might disappear for a while until he feels better.

It must be difficult for a mother cat if her kittens don't survive.

If one of her kittens is ill, she may feel sad but still destroy it to protect the other kittens. This is her means of self-preservation. However, if her instincts tell her it's not a crucial problem, she may just reject and abandon the kitten. But a total loss of a litter might seriously affect a mother cat's behavior.

Such as?

Sheba is a young cat whose people found her abandoned on a busy street with one kitten. The next day she delivered another kitten. Neither kitten survived. Within ten days she had a couple more, but they didn't survive either.

At first Sheba didn't react in any unusual way. Her people remarked that she appeared sad, but she was so thin and undernourished herself, she didn't have much energy to spare. As soon as her milk dried up and she put on some weight, she was spayed. Her people continued to keep her isolated, as they have an abundant number of cats. Fortunately, they also have a large house and yard to accommodate them.

Why was Sheba isolated to begin with?

Because a mother cat is usually very protective of her kittens, and Sheba's street experience intensified her need for privacy. When her people did try to introduce her to the other cats, her reaction was powerful! She wanted no part of them and would go into a rage whenever she even sensed a cat outside her door.

Sounds like she was more mad than sad.

Sheba's hostility was triggered by her loss. Whenever she had contact with another cat, she became anxious, relived her traumatic loss, and lashed out to protect herself.

You mean she felt threatened by the other cats?

Yes, she was still in a vulnerable state, and her sadness turned to rage.

Well, what was your advice to Sheba's people?

I recommended they give Sheba a lot of attention to build up her feeling of security, and start her on a tranquilizer to relieve her anxiety.

I never realized that cats' behavior was so indicative of the way they feel. You really have to be on top of things to know what your cat's trying to tell you.

Indeed so. The more you can interpret your cat's actions, the easier it is for you to understand how he feels. This understanding can only pave the way to increased harmony between you and your cat.

To make it easier to interpret your cat's feelings, I've created a chart to help you read his behavior.

How to Recognize Your Cat's Emotions

Emotion: Happiness
Stimulus: Stroking; cuddling; eating; nursing; close contact; snoozing alone or with a companion
Cat's Response: Face and body relax. There may be a smile on the face. Purring; stretching; kneading; licking

Emotion: Controlled Happiness
Stimulus: Qualified contact. Cat may purr as he sits in your lap but object when you try to touch him.
Cat's Response: Cat stops purring and reacting as above to glare at you. He may shift position away from stimulus. But if his tail flicks or his body ripples, it's his signal that he wants the stimulus to stop.

Emotion: Sadness
Stimulus: Sickness; loss of companion or person; feeling of neglect; separation anxiety
Cat's Response: Lethargy; dull fur coat; loss of appetite; bizarre toilet habits; scratching or pawing at various parts of the body or skin

Emotion: Rage
Stimulus: Presence of new or alien animal or person; accident; cat spat; tomcat or unspayed syndrome
Cat's Response: Body contracts. Rapid breathing; flattened ears; puffy tail; flicking tail; arched back; loud yowl. Cat may attack person or animal that was source of his anxiety or may displace his aggression to an innocent victim. Rage erupts quickly.

Emotion: Fleeting or Expectation Anxiety
Stimulus: Short-term stimulus (example: a fly on the loose that the cat wants to catch)
Cat's Response: Flexes body muscles; makes chutter-chutter noise.

Emotion: Prolonged Anxiety
Stimulus: Continued, unresolved anxiety; continued presence of an alien person or animal; tomcat or unspayed syndrome; presence of a sick animal, whether obviously ill or not.
Cat's Response: All of above responses may occur, or only a few plus medical problems may be triggered. Cat's appetite, litter box habits, disposition might be noticeably affected. Source may become a tension target.

Emotion: Overstimulation Anxiety
Stimulus: Too much arousal from being petted or held. Can't handle energy charge.
Cat's Response: Tail flicks, ears flatten, body ripples, cat may bite or scratch if action continues.

Emotion: Separation Anxiety
Stimulus: Temporary or permanent separation from person or companion.
Cat's Response: May become insecure, suffer anxiety and take it out on companion or person. Medical problems may be triggered.

Emotion: Eager Anticipation
Stimulus: The cat wants something from you (for example, to sit in your lap, have his head scratched, have the litter box changed).
Cat's Response: Stares at you; meows; jumps up beside you; rubs head against your hand; circles litter box or runs back and forth to it.

Emotion: Dread Anticipation
Stimulus: Your cat knows what's going to happen next and doesn't like it (example: grooming, traveling, ear cleaning).
Cat's Response: Cat disappears; body ripples; tail flicks. May pretend he's napping to delay action.

Emotion: Ambivalent Anticipation
Stimulus: Your cat wants and doesn't want something (example: wants to join party but is too shy).
Cat's Response: Encourages some petting but runs off quickly to discharge unacceptable high energy.

chapter six

choice and introduction of a companion for your cat

"Charcoal is ready for a kitten" was the message I received from The Cat Practice. The message was from Ellen, one of my clients, who had had to find another home for her cat, La Put. (See Chapter 1.) It was some months later, and Ellen had adopted a young street waif whom she had named Charcoal.

I called Ellen, and she told me how she and Charcoal had struck up a wonderful relationship. But she was away most of the day, and she felt Charcoal needed an adopted kitten to entertain her. "How old a kitten should I get, and does the sex matter?" asked Ellen.

I told Ellen that the kitten should be between eight and sixteen weeks old and that its sex was immaterial except for her preference. Ellen replied that she and Charcoal could easily accept a male or female, and asked me to be on the lookout for any homeless kittens. She also wanted to know if there were any particular instructions she should follow to get the introduction off to a good start.

This, of course, was a leading question! I told Ellen that the best thing she could do was to have a person unknown or little known to Charcoal deliver the new kitten. This way Charcoal wouldn't feel she was being betrayed by friends. Also, although I knew it would be difficult with a new, cute kitten

present, Ellen should devote all her attention to Charcoal. This would force the kitten to seek out Charcoal for contact and attention. I added that if Ellen couldn't ignore the kitten until Charcoal had accepted it, the adjustment would take longer. Ellen felt she could keep her distance from the new addition for long enough, and she would make sure that her boyfriend adhered to the same "hands-off" policy.

A few days later I called Ellen with cheery news: a homeless female kitten had turned up in the parking lot of the hospital where my husband, Paul, did surgery. Pauline, one of the nurses, was caring for it but wanted to place it in a permanent home as quickly as possible. I told Ellen that the kitten was about six weeks old. Ideally, eight weeks or older is a better age for a kitten to leave its mother. However, since there had been no mother cat on the scene, the kitten's fate had already been decided.

Ellen's response was terrific. "The kitten sounds perfect!" she cried and wanted to know how soon she could have it.

Within a couple of days Pauline had delivered the kitten, who was soon named Yenta, and Charcoal now had her new companion.

I wasn't surprised when Ellen telephoned to report that Charcoal wouldn't let Yenta near her, and that the kitten had spent part of the day behind the sofa. Ellen couldn't understand why Charcoal was such a bully. She was afraid that Charcoal might seriously hurt the kitten.

Although Charcoal's behavior appeared unreasonable to Ellen, it was natural, given the situation. Charcoal had to be sure that Yenta was not going to deprive her of Ellen's attention or any of her daily needs. Until Charcoal was satisfied that Yenta was not a threat, she would continue to ignore and even bully her. However, I pointed out that Charcoal wouldn't physically hurt Yenta. Her "remarks" would actually be more threatening than her actions, and if Ellen didn't weaken but kept her distance from Yenta, Charcoal would accept her that much more easily.

So many of my clients fear that their older cat will physically hurt a newly adopted kitten. But that is just another cat myth. True, the older cat will usually put on a fierce façade, which usually consists of loud hisses, growls, and tail swishes, to reinforce his superior position. However, the older cat will not hurt a kitten, provided you don't literally force the kitten on him

but instead leave the acceptance up to him. (The exceptions to this rule are a street tom protecting his territory and a mother cat who may destroy a sick kitten to protect herself and the rest of her litter.)

By the next week I had received a glowing note from Ellen. She wrote that Charcoal and her new kitten were getting on famously. The changes in Charcoal were wonderful. She was more relaxed, and her kitten gave her no time to get bored. Ellen added that if I ever needed a testimonial for adopting a companion kitten, she could go on nonstop.

Indeed, Charcoal and Yenta were off to a fine start, largely due to Ellen's handling of their introduction. (See section in this chapter entitled "How to Successfully Introduce a Kitten to Your Cat.")

You really believe that the people's behavior affects the relationship between cat and kitten?

There's no doubt in my mind.

But I had a two-year-old male cat, and I brought home a young male kitten, just dumped them together, and they became the best of friends.

I'm really happy to hear that. Your situation was very fortunate. Most introductions between animals, however, are affected by territoriality and sexual energy, especially if the older cat isn't spayed. They often turn into hissing and yowling sessions—and worse. Evidently you paid enough attention to your cat, and his disposition was mellow enough, so there weren't any major problems.

What do people usually do wrong when they adopt a kitten for their older cat?

The most common mistake they make is to indulge the kitten with all kinds of attention and spend less time with their older cat. They don't realize that they're giving their older cat less time, but he does! He feels the difference very strongly, and this usually causes him to reject the new kitten, and in some instances to reject his people.

Why would he do that?

Because he feels neglected, and his feeling can turn to anger and/or unhappiness. When this occurs, the cat may respond by withdrawing or becoming grouchy and/or aggressive.

What's the best thing to do if this occurs?

Pay your cat double attention. Ignore his pleas to "get away, don't touch me!" Do all you can to draw him out. This is the time to reassure him that he's top cat and that the kitten is for him. If your cat feels that the kitten is for you, he feels threatened.

But how can you make a cat feel an adopted kitten is for him? You can't expect him to understand what you tell him.

Although your cat may not understand your words, he responds when his name is mentioned and when he gets the feeling that he's being included. You have to tell him that the kitten is his new friend, and take pains to include him in the conversation whenever you refer to the kitten.

Say things like, "He's not a bad kitten, but he'll be even better after you teach him how to do things." But remember to say this or whatever else you choose to say in a sincere tone. Your cat can tell how sincere you are by the tone of your voice and your body language. It's hard to hide your feelings from him.

Any other pointers?

Actions speak louder than words. Be sure not to cuddle the new kitten for some time, in or out of your cat's presence. The more you touch the kitten, the less your cat will, and the longer he'll take to accept the kitten. Remember, he can smell your touch on the kitten's fur.

But if my cat and I ignore the kitten, won't the kitten feel deprived?

If you ignore the kitten, his feeling of deprivation will force him to seek attention from your cat. Kittens are very persistent and fearless, and these specific qualities will eventually win your cat over.

I see. If I provide the kitten with attention, he'll be more reluctant to seek out my cat.

Absolutely! Why should he bother to exert himself in winning over your cat if you're giving him all the attention he needs? A kitten is so adaptable, tireless, and persistent that winning a cat friend over is hardly an effort for him. It's all part of being a kitten!

How long does it usually take for the two to become friends?

Anywhere from a few days to a few weeks. It depends on how malleable the cat's personality is, the kitten's perseverance, and the people's noninterference.

Do you know of any cases in which it took longer or where the older cat wouldn't accept the kitten?

Yes. I'm particularly reminded of a cat named Iago and his adopted kitten, Oberon. His people adopted the kitten to keep Iago company, but when Iago ignored Oberon, they filled the gap. To complicate matters, soon after the adoption they left on a holiday, and the cats were looked after by the cleaning lady. Consequently Iago stayed clear of Oberon, rejected his people, and spent most of his day on top of an out-of-the-way cabinet.

The people suspected that their cleaning lady probably also indulged Oberon with affection, as Iago is less likely to approach people—he is not people oriented. I explained to them that even though Iago didn't ask for affection like Oberon, he would still be bothered if Oberon was getting lots

and he wasn't getting his fair share. That's why when his people returned, Iago withdrew from everyone and hid on top of the cabinet.

So what did you recommend?

I told the people that they must do their best to make Iago feel good. Even though they had gotten off to a bad start, it wasn't too late. They would have to stop showing the kitten any affection until Iago accepted him as a friend. At first Oberon would be a little confused, but then he'd switch his attention to Iago.

How could they tell when they could pet the kitten?

They saw Iago begin to interact with the kitten. He allowed the kitten to approach him and hang out with him, without taking flight at every encounter. Then they started to pet the kitten; but when they did, they would mention Iago's name and frequently include him in the petting.

Did it work out?

Yes. For a while it was difficult—after all, kittens are hard to resist and cats are hard to win over—but Iago's people did a superb job of making him feel good. Soon, Iago left his perch on top of the cabinet and became the cat he used to be. The best part was when people noticed that Iago allowed his kitten, Oberon, to play with his tail and romp alongside him.

What else would you have suggested if Iago hadn't responded?

Than Iago would be started on tranquilizers or anti-anxiety drugs to remove his anxiety.

Suppose Iago still didn't accept the kitten even with the help of a tranquilizer, what then?

I would have told his people that they'd have to find a new home for the kitten. It would be easier for Oberon to adapt to a new home while he was still a kitten. In time, they could try Iago with a new adopted kitten. With the help of my special method of introduction (given later in this chapter) and their newly acquired cat experience, they would succeed.

I have a young kitten, and I'd like to adopt a one-year-old cat. Do I devote most of my attention to the kitten, since the cat will be the newcomer?

No, even though the cat will be the newcomer, he's the one who needs the most support.

Why is this?

A kitten is more adaptable and won't feel threatened if you pay attention to his new companion. However, because it's already the kitten's territory and the cat is the newcomer, you can still interact with the kitten.

What about if you want to adopt a young cat and you already have a young cat?

It's your present cat whom you dote on. The acceptance will probably take longer than if you were introducing a kitten. But if the two cats are of mellow disposition, the outcome should be positive. Make sure that both cats are already neutered if they are sexually mature. Wait at least two weeks after the neutering for the hormones to change before you schedule the introduction.

Why do you think that a cat objects to the presence of a new companion in his household?

A cat is a very territorial creature. If a new kitten or cat appears in his territory, he is threatened because the newcomer represents less for him. All of a sudden, here's this newcomer sharing his litter box, his food, toys, favorite spots, and if he's not careful, the newcomer will take over everything.

Including his person?

Yes, that's why it's important to ignore your cat's new friend until the two have started a positive relationship. If your cat feels he's going to be denied your attention, and you reinforce his fears by petting or holding the newcomer, you can be sure he'll harbor resentment.

That's why he'll resent his new friend and even take it out on his person?

Yes, indeed. So why breed jealousy when there's a way around it!

Do you really think that every cat needs a kitten?

I think most single cats benefit by having a kitten or cat companion. This is especially true of cats with high energy; they become victims of the single-cat syndrome.

What is the single-cat syndrome?

That's what I've dubbed misbehavior that cats indulge in out of boredom and loneliness. Being deprived of an animal companion seems to emphasize a cat's destructive and/or aggressive characteristics. His acts are primarily attention-getting devices, such as destroying the furniture, wrecking the ornaments, and/or displaying rough or hostile behavior to his people or visitors.

You mean a kitten would help cure this syndrome?

Yes, because the cat would transfer his energy to the kitten—and kittens have plenty of energy to keep the cat occupied and prevent him from resorting to negative outlets.

But suppose the cat's person can provide him with abundant attention?

That's fine, but even so most cats need to interact on a cat-to-cat or cat-to-dog basis to attain a healthy and contented release. They just can't tussle with a human being in the same way, or chase and leap together as cats do.

Still, aren't there some cats that do best being the only cat?

I have encountered some. A friend of mine had two cats, and when Schroeder, his older cat, died, he adopted a young male kitten for his remaining cat, Hobbit. Although Hobbit had enjoyed a fine relationship with Schroeder, she was not pleased with her new companion. Several weeks later, in spite of my friend's endeavors, Hobbit continued to ignore the kitten. Then, as it happened, someone else fell in love with the kitten, and the kitten moved on. My friend realized Hobbit was happiest without the kitten. Perhaps this wasn't the kitten for her, or maybe she had decided that at nine years of age she was too old to take on the responsibility of teaching a kitten. Or perhaps she chose to remain true to Schroeder and preferred to remain a widow.

I remember another cat that appeared to prefer a solo performance. Some years ago, after appearing on a radio show with musician/composer Barry Gray, I received a message to call his wife, Judith. Their cat Nichevo had recently been joined by two young kittens. Several months before, Nichevo had lost his beloved companion to a critical illness. A stray female waif whom he feared was also in temporary residence.

Judith and Barry were worried because Nichevo tolerated but didn't exactly reach out to the kittens, and his bowel movements were irregular. They loved Nichevo and hoped he hadn't been affected by the two kittens.

I explained that, as they knew, Nichevo is a very sensitive cat, and the loss of his companion bereaved him. Furthermore, the presence of the street waif threatened him, and his stress tolerance was lowered. The two kittens presented a doubly high energy level and created a triangle that made it difficult for Nichevo to integrate them into the family, as he was easily intimidated. While I talked, Nichevo appeared from one room as the kitten retreated to another. He wanted his time alone! As he kneaded Judith's stomach, I could see by his face and body just how fragile and delicate he was.

I recommended that Judith and Barry continue to draw Nichevo out and consider starting him on a tranquilizer to relieve his anxiety and relax his skeletal muscles. This would also aid in relaxing his bowels, releasing his body tension, and helping his regularity. I added that it might be a wise diagnostic measure to have his chest and heart X-rayed to rule out an underlying medical problem (perhaps a sonogram or ultrasound). It would take some time, but Nichevo would respond more comfortably to the kittens the more he felt Judith and Barry loved and reached out to him.

What kind of things do cats teach to kittens?

Cats teach kittens the cozy spots in which to hang out; when they should retreat from their people's moods; how to scratch the scratching post; how to gain affection from their people; and countless other valuable cat endeavors. It's really a joy watching a kitten copying his cat.

Any other situations in which you don't recommend introducing another cat?

Generally, if a cat is over ten years old, it's difficult to introduce a new companion. This applies especially to a cat that has medical problems and/or tends to be more people oriented than cat oriented.

What do you mean by people oriented and cat oriented?

A people-oriented cat is a cat who, although he interacts well enough with other cats, still relates best to people. A cat-oriented cat is one who usually is wary of people and prefers the company of other cats.

Do you really think a cat can prefer one kitten to another?

Generally any young kitten will do, but there are many exceptions. If possible, it's best to match your cat's "catsonality" when you decide to adopt a kitten.

How would I do this?

Choose a cat-oriented kitten if your cat is introverted and shy. This type of kitten would be shy of people contact but play and interact happily with her littermates and other cats. A cat-oriented kitten would naturally first strike up a relationship with your cat before it made any overtures to you. This behavior would speed up your cat's acceptance.

Any other reasons why a cat may choose one kitten over another?

Color is another consideration. A cat is sensitive to the color of other cats, perhaps because cats of different colors have distinctive smells; and cats are also sensitive to various color patterns. I've known several cats that invariably have aversions to cats of a particular color. For example, my cat Sam used to react violently to orange cats. Cary Grant, an ex-resident of The Cat Practice, had the same reaction. They just couldn't deal with redheads!

How do I figure out what color kitten my cat would like?

Observe your cat's interactions with other cats as much as possible. If there is a mother with a litter, you will notice the mother favors certain color kittens (often her own color) and you will see that certain kittens play together. Other than that, try to pick a color kitten that matches or harmonizes nicely with your present cat's color, and hope your cat agrees with you!

I've devised a set of personality pointers to help you find the perfect mate for your cat; why take a chance and introduce a random type of kitten! You can ensure a happier relationship if you consider your present cat's personality:

- Greta Garbo type: If your cat's the Greta Garbo type, not terribly outgoing, adopt a spunky but calm kitten so your present cat doesn't go off "to be alone." A cat-oriented kitten—one that prefers cats to people—would be best, so the kitten will reach out to your cat before befriending you.
- Mickey Rooney type: If your cat's the Mickey Rooney type—Mr. Personality Plus—a bouncy, fearless kitten would do fine.
- Gary Cooper type: If your cat is mellow and even-tempered, any type kitten would be suitable.
- Bob Hope type: If your cat has a protective motherly or fatherly nature, you could adopt even a tiny and timid kitten.
- Marilyn Monroe type: If your cat is very glamorous and pleased with itself, be sure to get a kitten cute enough to hold its own in the beauty contest.

General Pointers for Matching Cats Happily

1. Don't adopt an Elizabeth Taylor or Clark Gable type cat if your cat is not equally as dashing. You don't want the newcomer to steal your cat's meow!
2. If you know that your cat has an aversion to cats of a particular color, be sure not to adopt a kitten of that coloring or markings. You don't want your cat to start off with a negative association.
3. Adopt a male kitten if your cat is a female and over four years old. If your cat has had a poor relationship with a particular gender of cat, adopt a kitten of the other sex. Otherwise, adopt whichever gender of kitten appeals to you.

4. If either your present cat or the adoptee is sexually mature, arrange the introduction at least two weeks after neutering.

How to Successfully Introduce a Kitten to Your Cat

The instructions that follow grew out of techniques I developed and tested at The Cat Practice. There is a method for introducing a kitten or cat to your cat, and a separate method for introducing a dog to your cat. It may seem like a lot of trouble, but the initial bother is well worth the lasting happiness that results!

Of my various methods of introducing a new kitten to a cat, this is the one I prefer. It works quickly and effectively.

Preparations for the Newcomer's Arrival

1. Remind yourself that the kitten is for your cat. After they have bonded, you can interact with the kitten; but continue to refer to it as your cat's kitten so the cat feels included and in control. However, to be sure their relationship is solid, wait at least ten days after they've bonded before you interact with your cat's kitten.
2. Remember that your cat had all of your attention before the kitten arrived. You want to avoid making him jealous. Your cat must not feel that the kitten is going to deprive him of your attention.
3. If he feels the new kitten is clearly his friend, they'll bond very quickly. Tell-tale signs: they engage in play together, groom each other, and exude happiness.
4. Schedule the introduction for a time when you'll be in good spirits. Generally, the morning or early afternoon is best for the introduction.
5. Treat both you and your cat to your favorite breakfast and/or lunch.
6. Don't plan any gala parties or renovations that day.
7. Provide an extra bowl for the newcomer. An extra litter box is usually not needed.
8. Praise your cat frequently and give him extra strokes and hugs.

The Newcomer's Escort

1. Arrange for someone to act as escort for your cat's new kitten.
2. The escort, if possible, should be someone your cat hasn't befriended.
3. The escort is needed only to escort the newcomer in, not to stay on.

The Newcomer's Arrival

1. While you and your cat are out of sight in the bedroom with your cat's favorite things, and the bedroom door closed, the escort should deposit the newcomer in the bathroom and open the lid of the carrier.
2. The next and final step for the escort is to leave the bathroom door slightly ajar and to depart.
3 When you hear the entrance door close, open the bedroom door so your cat can slip out and eventually discover his "new friend."
4. If you're too curious to remain in the bedroom, the best thing you can do is to leave home and entertain yourself elsewhere for at least a few hours. It would be best to leave when the escort leaves so your cat can work it out without your interference.
5. Your cat can take care of himself and his new friend.
6. When you do return, remember the kitten is invisible, you only have eyes for your cat. If the kitten jumps in your lap or tries to seduce you,

remember he's invisible. The more you interfere, the longer it will take for the two to bond. You must be brave! If you ignore the kitten, he will seek your cat out over and over again until your cat realizes that he's still top cat and that the kitten's welfare is his responsibility.

7. At feeding time, tell your cat you're feeding his kitten and that his own food is safe.
8. Whenever you do something for the kitten, mention your cat's name so he feels included—even after they are buddies. Tell your cat you're taking care of the kitten so he doesn't have to do it.

If it's impossible for you to have an escort do the introduction honors, here's how you can do it yourself.

Alternative to Escort

1. Leave the newcomer in his carrier outside your door so your cat doesn't catch you cat-handed!
2. Greet your cat as usual, wash your hands, and busy your cat with a snack. Be as relaxed as you can, and don't hold your breath!
3. Quickly but calmly, while your cat is occupied, open the front door, take the newcomer from the carrier, and place him inside the door.
4. If time permits, place some strips of newspaper, etc., from the inside of the carrier where your cat can find them. This will familiarize him with the newcomer's smell. Be sure to wash your hands so your cat doesn't think you're cheating on him.
5. Don't monitor their every move. Try to engage yourself in a constructive but non-anxious project.
6. It is all right to leave them alone if you have to go out. My preferred method of introduction is for you, the guardian, to leave so they can work it out without your input and fears.

REMEMBER: A cat will not hurt a kitten (unless it's sick or you provoke your cat's defenses). Action speaks louder than hisses, which may occur frequently. Relax, and they will, too. Don't try to manipulate them. Generally, a cat is repelled by such action. If the newcomer is an older cat who's being introduced to a kitten, give the newcomer attention and refer to your kitten as the cat's friend. The kitten is more flexible and will seek out the newcomer for the attention he lacks from you.

IMPORTANT: The kitten is also "invisible" to your friends. They, too, must give all their attention to your cat until it's plainly apparent that your cat has accepted the kitten without reservation.

I frequently make the introductions. My expertise is especially needed when a particular cat has had a poor track record with previous cat relationships or has always been a single cat. In these introductions, I assume the role of the escort. However, unlike the escort, I remain to monitor and assist with my appropriate relaxation techniques and the guardian can remain because I am there.

I once used this method to introduce a new kitten to a ten-year-old Himalayan named Jamara who had multiple chronic health problems. Their bond has given Jamara a new and happy lease on life.

I made another auspicious introduction—a two-year-old male to an eight-year-old cat named Nantucket who had recently lost his nineteen-year-old senior companion. Within five days Nantucket and his new friend played and slept together.

Yet another success story—Jesse, a two-year-old cat, discovered his new adopted kitten on the edge of the bathtub and greeted him with a lick.

If you keep clearly in mind, and carry out in practice, the policy that the newcomer is your cat's priority and your cat is yours, harmony will prevail rapidly. If your cat seems to ignore you while he cultivates his new relationship, don't fret, because it's only temporary and not catastrophic. You'll very shortly be reunited and delighted!

Remember, the key word is **invisible**!

Do you think it's a good idea to introduce a third cat into a household?

I generally advise people that, as with human beings, three cats make a triangle situation that generally leaves one cat stranded and feeling very rejected.

But what if one cat likes to play and the other likes to be left alone?

Unless you are sure that a newcomer won't upset the status quo, it's best not to adopt a third.

What's your feeling about adopting a kitten to live with a dog?

Sounds fine. But if you have a large dog, adopt a kitten that's at least four months old in order to avoid accidental physical injuries.

How to Introduce a New Cat or Kitten to a Dog

Adopting a kitten or cat for your dog or puppy should be easier than introducing the latter to the former. A dog is a pack animal and generally takes well to direction, and a kitten or cat will supply exactly such inspiration.

Matching Your Dog's or Puppy's Dogsonality

- If you decide on a kitten, it should be at least three months old in order to prevent any playful accidents caused by the puppy or dog.
- Adopt a mellow cat or kitten if your dog or puppy is shy and/or reclusive. The same applies if your dog is over seven years old.
- Don't adopt a shy cat or kitten if your dog or puppy has a high energy level.
- If either the cat or the dog is sexually mature, arrange for neutering at least two weeks before the introduction.
- Don't adopt a type of kitten or cat with which your dog or puppy has had a poor experience.
- Select the gender of cat that your dog prefers, if the dog has a preference.

Introducing a New Cat or Kitten to Your Dog

1. Give your dog an object with a cat or kitten scent so he'll become accustomed to the new smell.
2. Set up a high spot or counter where the newcomer can eat or just perch without the dog's or puppy's interference.

3. Purchase or build a scratching post for the newcomer with the essentials I have listed in Chapter 3. A kitty screen might be handy if the litter box is very accessible.
4. Make sure the newcomer's nails are trimmed.

Cat or Kitten Day

1. You can introduce the newcomer. Since the canine species responds well to authority and direction, your dog or puppy will more easily adapt to the newcomer's appearance if you're the go-between.
2. The cat or kitten should be in a see-through carrier. Place it in a spot where both animals can observe each other but at a healthy distance.
3. Provide the newcomer with an object that has your puppy's or dog's scent.
4. After a short interval, open the newcomer's carrier in the room where you will set up the litter box. Close the door so your puppy or dog can't gain entry.
5. Use a screen or gate to separate the two from each other.
6. Be sure to put the newcomer's belongings with him.
7. After a few days put each pet in the other's territory, so they'll become better acquainted with each other's smell.
8. The newcomer may decide to cross the gate or screen to join his new companion. Leave it up to the newcomer. Never force the issue.
9. It is all right to have contact with the cat or kitten even before he's accepted. However, don't omit attention to your dog or puppy—remember he was your first companion. Don't let him take his jealousy out on the newcomer. Mention the dog's name whenever you touch the newcomer, so the dog feels included.

Encounter Day

1. If the two haven't already made contact, bring about the official encounter on a day when you are feeling happy and relaxed.
2. Remove the gate or screen so they can have easy access to each other.
3. Don't interfere! It's up to them!
4. Disperse them with a plant sprayer if they get too excited.

5. Stay calm and collected. Don't transfer your anxiety to them.
6. When you leave them alone, replace the gate or screen. Continue to do this until you're sure of their actions.

Generalities

1. Sedation may be required if either animal's anxiety level is high.
2. Give the newcomer attention, but don't forget to have close contact with your puppy or dog.

Grand Finale

You should have a loving or at least tolerant couple within three weeks.

I have a hunting dog and live in the country, and I'd like to adopt a cat or kitten.

Unless you're positive that your dog is not a cat chaser, I'd think twice about adopting a cat. The relationship between hunting dogs and cats can work out, but it will take some time and patience. Generally, it's easier for a dog to accept and interact well with a cat within the dog's own environment, but a hunting dog's temperament may complicate matters. Because he is bred to hunt other animals, you would have to keep him closely in line.

Would it be best to adopt a kitten or a cat for my dog?

It would be easier to introduce a kitten to your dog, because a kitten is more adaptable than a cat: he doesn't have any previous dog associations. But if you'd prefer a cat, adopt one that you know has had positive experiences with a dog. However, if the cat is sexually mature, be sure to have him neutered at least two weeks before the introduction.

I have a seven-year-old cat named Spunky. Although she loved the two dogs we used to have, she won't accept our new female German shepherd puppy. She spends most of her time upstairs by herself. Our other dogs died of sickness and old age.

Spunky may still be grieving for her friends and may not be used to the high energy level of a puppy. Do you have any other animals?

Yes, I have three other cats. She gets along well enough with them.

I would make it a point to give Spunky more attention to make her feel happy. Probably the puppy will calm down as she matures, and Spunky will be able to accept her. In the meantime, however, if you feel Spunky is too depressed, it might be good to tranquilize her for a while. The tranquilizer will relieve her anxiety both emotionally and physically. When you notice a measurable improvement in Spunky's behavior, you can slowly lower the tranquilizer dose until she is off it.

I thought cats and dogs were natural enemies.

When they are raised together, a cat and a dog usually become good friends or at least work out a tolerant relationship. However, the same cat or dog may very well act aggressively with other cats or dogs—especially outdoors.

Who's the dominant one in a cat-and-dog relationship?

In many instances it's the cat who takes the lead. It's usually because a dog responds well to instructions and a cat makes no bones about exerting his authority. For example, I know of a large Irish setter whose family adopted a young kitten named Yeti. Although the setter carries the weight in the family, it's Yeti who bosses the setter.

How do you feel about introducing a dog into a cat household?

Actually, it depends upon the "disposition" of each. It would be easier to bring a dog into a kitten household than it would be to introduce a dog into a cat household.

Any particular pointers?

Try to match. Don't adopt a high-energy puppy or dog if your cat or kitten is shy and introverted. A mellow disposition is your answer.

I have an eight-year-old cat who has never lived with any other animals. Would it be best for me to adopt a dog or a puppy to be her companion, and what about the sex?

I would recommend that you adopt a dog, rather than a puppy, and if possible one who has lived with cats. A male would be best for your cat; your female would find a male less competitive, so she'd feel less threatened.

If you want to adopt a puppy, choose one that is at least a few months old and appears to have a calm disposition. (See "Matching Your Cat's Catsonality" on the next page.)

My aunt has a two-year-old neutered male cat and a six-month-old female dog. They have a terrific relationship, but lately the cat seems a bit nervous and is urinating around the house. What's wrong with him?

Sounds like the dog is approaching sexual maturity and is about to reach her first heat. The cat has become sensitive and anxious about the dog's new status.

I would recommend that your aunt have her dog spayed after the dog goes out of heat and also have the vet check out the cat. The cat may need medication for his bladder and a tranquilizer to relieve his anxiety until the dog's heat is over. Fortunately, a dog usually goes into heat only twice a year.

What's your feeling about introducing a bird or hamster to a cat?

I definitely suggest you take the utmost caution, so that your cat can't harm the other animal. If it is a bird, hang the cage from a very high, secure post. A hamster should be kept in a closed environment so that your cat can't gain access.

Can a cat actually interact with a bird or rodent without attacking?

Yes, but this takes time, patience, and good sense on his person's part, because a cat instinctively will treat such companions as prey.

Generally speaking, a cat or dog makes the best companion for another cat. There may be upsets at the beginning of the new relationship, but it will be worth the companionship and love that both you and your cat will gain.

How to Introduce a New Dog to Your Cat

If you decide that there must be a dog or puppy in your cat's life, there is a special procedure you can follow to ensure you get the two off to a good start.

Matching Your Cat's Catsonality

1. Adopt an older male dog with a mellow, placid disposition if your female cat is over four years old and not dog oriented, so she won't be intimidated.
2. If you have a male cat over four years old who is not dog oriented, adopt an older female dog with a mellow personality.
3. A young puppy would be suitable if your cat's very outgoing and/or has interacted favorably with dogs.
4. A calm puppy or dog would be best for a timid, young cat. The sex of the dog is up to you.
5. If your cat has had a good relationship with a particular kind of dog, try to select the same kind. Conversely, don't adopt a type of puppy or dog with whom your cat has experienced a poor relationship.
6. If either the cat or the dog is sexually mature, postpone the introduction until at least two weeks after neutering.

Matching Your Kitten's Catsonality

1. To avoid accidental injuries from the puppy or dog, wait until your kitten is at least three months old.
2. Adopt a calm, relaxed puppy or dog if your kitten is shy and not used to dogs.
3. If your kitten's the friendly sort (personality plus) and/or a very mellow fellow, any dog or puppy should do.
4. If your kitten has had a traumatic experience with a dog or puppy, I would not plunge into adopting another.
5. Sometimes it is easier to start with a kitten and puppy so they can grow up together.

Preparations for the Arrival of the Dog or Puppy

1. Prepare your cat or kitten for the newcomer by letting him listen to a barking dog on tape, CD, or computer. Don't play it too long the first few times; let him get accustomed to the new sound slowly. Be careful not to blast the volume, because a cat is very sensitive to sound. Play the sounds when your cat is eating so that he forms a positive association with the barking: You want your cat or kitten to experience the sound without fear or apprehension.
2. Change your cat's or kitten's feeding place to a high spot so the newcomer doesn't invade the food. Don't set up a territorial tug of war. Your cat or kitten should have a high retreat where he can escape from the newcomer's antics when necessary.
3. Trim your cat's or kitten's nails.
4. It might be advisable to purchase or build a kitty privacy screen (see pages 55–56) if the litter box is easily accessible to the newcomer.
5. Make sure that a scratching post is available and in an area secure from the curious newcomer. Add some extra catnip to entice your cat to scratch. It's a good way to expend increased anxiety and create positive feelings in the cat or kitten.
6. Tell your cat or kitten that he is soon going to have a sturdy protector, but it's up to him to teach the newcomer some cat wisdom.

Dog or Puppy Day

1. The newcomer should be introduced by an escort who is unfamiliar to your cat or kitten. If you act as the escort, your kitten or cat may feel you have betrayed him.
2. Keep the newcomer in a separate room with all of his belongings for the first few days. If he has to be walked, try to keep him near the exit door. (Keep him in a crate or on a leash if you can't keep him separate.)
3. Use a screen or gate to separate the room, one the newcomer can't jump over, so they can see and smell each other.

4. If the newcomer is old enough to go outside for walks, try to arrange for a neutral party to walk him for the first few days. The less contact you have with the newcomer, the quicker your kitten or cat will accept him. Give your cat or kitten a treat before the newcomer's walk if you must be the walker. Tell him you're coming right back.
5. Try to keep the newcomer from becoming excited, so his high energy is not a threat to your cat or kitten.
6. After a few days have passed, switch territories for about an hour so that each can get a good whiff of the other. Repeat this for a few days, until they have become less territorial.

The Friendly Encounter

1. Pick a day when you're feeling relaxed and comfortable about the encounter.
2. Take away the gate or screen so they can confront each other if they wish.
3. Remember to let them proceed at their own speed; the less interference from you, the better.
4. Use a plant sprayer to separate them if they get too excited.
5. When they are out of your sight, return them to their initial territories to avoid any possible problems. Continue to do this until you're confident of their actions.
6. Do your best to stay relaxed, so that your anxiety isn't transferred.
7. Try to have other people provide the newcomer with attention, if possible. The less contact you have with the newcomer, the less threatened your cat or kitten will be, and the more quickly acceptance will come.

Generalities

1. It may not be necessary to be so cautious if you're introducing the newcomer to your kitten. A kitten is usually less fixed and more flexible than a cat.
2. Don't skimp on the attention you give your cat or kitten.

3. Sometimes it may be necessary to use sedation for the newcomer and/or your cat, if the anxiety level is high.
4. If your cat goes outdoors, he may accept the newcomer faster because he's less confined.
5. Many times, as a newcomer a dog is less of a threat to your cat than another cat would be.

Grand Finale

Your cat or kitten should work out a living arrangement with the newcomer within two weeks at most. The relationship may become one of mere tolerance or may blossom into one of genuine camaraderie. Don't be surprised to see them sleeping together or taking turns washing each other. If your cat goes outdoors, you may find that the cat will choose to accompany the newcomer on his walks. There may be upsets at first, but it is worth the love and companionship both you and your cat will gain.

chapter seven

choosing the best diet

A cat's diet directly affects his physical and emotional state. If his diet is not well balanced, his health may suffer. Beau and Prudence clearly illustrated this premise. Their person, Harry, was a great believer in the cat–tuna myth. He had been influenced by advertising claims that cats thrive on fish and that tuna is their favorite. These claims are right in one respect: Tuna is a flavor that cats enjoy.

"But tuna has always been all that Beau and Prudence would eat," Harry sadly replied. "Do you really think that's why they have skin problems and why Beau beats up Prudence so often? He has gotten so hostile he even lashes out at me!

"I explained to Harry that "tuna-fish junkies" almost invariably display these distressing symptoms. Tuna has been known to, possibly, cause a vitamin E deficiency in cats, manifesting itself in skin problems, frequent loss of appetite, lethargy, and nervous and/or aggressive behaviors.

"But how can I get them off tuna?" Harry asked wistfully. I told Harry that he could gradually add either canned or cooked chicken or beef to the tuna. It would take a while—from four to six weeks—but tuna would gradually be eliminated from their diet. I also recommended that he consult his vet about medication to stimulate their appetites and to treat their skin.

Although it was a long haul, Beau and Prudence were cured of their addiction. Harry found it a difficult time because for several days they rejected almost all their new food and picked out the tuna only. At times he confessed he was worried that they would starve! But he reminded himself that they weren't lean and could live off their body resources until they accepted their new food regimen. Five weeks later their tuna-fish addiction was gone. The positive improvement in skin and temperament reflected their new diet, and Harry was pleased by their renewed affection.

Although your cat may show a strong or exclusive preference for fish, you can't be guided by his choice. Many people, including myself, would adore an exclusive diet of ice cream and cookies. It is up to you to select a well-balanced diet for your cat.

What do you consider a well-balanced diet?

Beef, beef by-products, and poultry—cooked or canned—should be your cat's primary diet, although some cats will reject canned chicken or turkey. If you feed your cat canned food, read the list of ingredients on the label carefully. Meat and meat by-products should be avoided because they often

contain horsemeat or pork, to which some cats are allergic. (To check, contact the manufacturer.) However, beef by-products are strictly beef.

Canned or cooked organ meats, such as liver, kidney, and heart, should comprise no more than one third of the weekly diet, because they do not contain all essential nutrients. Raw meat has always been a controversial issue, but there are cats who thrive on it.

Fish should be limited to one fourth or less of the weekly diet. Avoid tuna, as already mentioned, because it is known to cause a vitamin E deficiency in cats that can trigger skin, urinary, and nervous/aggressive disorders. Many cats become addicted to tuna, and it is difficult to wean a tuna junkie. Also, if your cat starts holding out for fish, eliminate it completely from his diet so he doesn't become addicted to it.

Although dry and semimoist foods are well balanced, limit your cat's intake to no more than one third of his daily diet. True, some cats do well on an exclusive diet of dry and/or moist foods, but the majority of cats thrive, both coat-wise and catsonality-wise, on canned and/or other people foods. If your cat's a little chubby, keep in mind that these foods are high in calories.

Your cat may enjoy meat, poultry, and/or vegetable baby food. Baby food can be a fine supplement, but don't feed him an exclusive diet of it (check the brand of baby food with your vet). Cooked and raw vegetables are also fine for your cat. However, don't try to make a vegetarian of your cat unless you have a "cat-proof" diet, because cats are natural carnivores.

Cheese and yogurt are good foods for cats, and they make tasty snacks. A few pats of butter a week are good for your cat's regularity and help his furballs pass through. If your cat rejects butter, try commercial gels like Laxatone and Petromalt or a powder called Siblin.

Some cats have food passions—they may adore melon and pitted olives. It really makes them feel good. If you find that your cat has a passion for foods such as spaghetti, pizza, ice cream, Twinkies, etc., don't deny him. But be sure also not to overindulge him.

If your cat is a finicky eater, you may have to be creative about stimulating his appetite. (Refer to hints at the end of this chapter.)

What about organic cat foods in which the ingredients read like human food?

Yes, they are very popular with cats, and are frequently better products. Again, check my website.

How can I find out what is in a particular cat food?

You must contact the manufacturer of the food if the label does not specify the contents.

Can cats thrive on a vegetarian diet?

Although there are experts who have worked out special feline vegetarian diets that certain cats have thrived on, cats are basically carnivores. You have to be very careful and exact in regulating an exclusively vegetable diet. Personally, I feel vegetables are fine if your cat enjoys them. They're a great low-calorie supplement.

What about cooking for my cat?

There are special recipes that contain the necessary nutrients for a cat's daily diet.

Are there supplementary foods other than those you have already mentioned?

Cooked chicken necks and backs provide good gum exercise. Semi-moist and dry food can be given in small amounts or as snacks. There are dry foods available that are a particular aid to healthy teeth (check with your vet if you have any questions.)

Does dry food cause urinary problems in cats?

If a cat has a urinary problem or a predisposition to one, it's best to rule out dry food unless it's a prescription or special formula, because it can aggravate the condition in some cats. In any case, limit the amount of dry and semimoist food, because some cats become addicted to them to the exclusion of all other foods.

Are there any vitamins you recommend?

You can add one-quarter to one-half teaspoon of brewer's yeast and wheat germ to the food. These food additives contribute to glossy fur. The yeast is high in the B vitamins, and it helps lower your cat's nervous stress. But mix these in well, so your cat doesn't get a mouthful of yeast. Some cats are brewer's yeast pill fans and will literally beg their people to give them another pill. Again, limit your cat's pill intake if he's on the chubby side. Your vet may also have some vitamins to recommends specifically to your cat. (Avoid brewer's yeast if your cat has a kidney or bladder problem.)

What about milk?

Milk is fine if it doesn't give your cat diarrhea. Cream doesn't have this effect on your cat's system, but may be hard on your finances. Half-and-half is sometimes the happy medium. Yogurt is another healthy dairy treat, and fresh water should always be available.

What if my cat's not a drinker?

Invariably, he'll get enough liquid from his food. You might salt his food to encourage him to drink more. Some cats prefer to lap their water from the tub drain; others are less aesthetic and choose the toilet bowl. Felix was a young cat who dashed to the sound of a running water tap. He was in ecstasy when the kitchen tap was faulty and the water spurted out as though from a drinking fountain. Your cat might have fun with a water fountain that's designed especially for cats.

How many daily feedings are necessary?

A kitten should be fed four times a day until he is five months old, because he has a small stomach. Fresh water should always be available. Then three daily feedings are sufficient until the cat is six months old. After that he should be fed twice a day.

A kitten's daily food intake should equal about nine ounces. From six months to a year old he needs seven or eight ounces a day. After that, five or six ounces a day will suffice. However, let your cat's figure be your rule.

Be sure to wash out your cat's bowls after each feeding. You might want to put a place mat under his bowls if he tends to be a messy eater.

Your cat's diet should consist mainly of protein and fat. (It is different for senior cats, see Chapter 16.) Most cats cannot digest carbohydrates very well.

I understand why kittens must eat several times a day, but why can't a cat be fed once a day?

A single daily meal can cause digestive problems, especially for cats who "inhale" their food. The digestive tract can be overworked, and in severe cases such quick eating can cause vomiting. Decreasing the amount of food given at any single time limits this problem. Also, if your cat's a nibbler, periodic feedings are best, to keep the food from being dried up and wasted.

Speaking of kittens—at what age can they be weaned from their mother?

Not until they are at least six to eight weeks old.

What if the kittens are reluctant to be weaned?

Guide them over to the plate of food and stroke and talk to them while they eat to help them feel content.

Can they eat anything in addition before then?

You can offer them baby food or canned food at four or five weeks.

Can the mother's milk dry up before then?

The mother's milk depends upon her health and food intake. If either is inadequate, her breasts will be barren. That is why you want to be sure to feed an expectant mother cat at least eight ounces of food a day (mixture of canned, cooked, and dry foods) and give her brewer's yeast and wheat germ supplements. Add one-quarter to one-half teaspoon of each to her meals, and mix in well. If she takes to it, add more. You might want to have her blood tested by the vet to see if she is lacking any vitamins or minerals that may need to be supplemented.

What happens if the mother's milk dries up early in spite of a good diet?

The kittens' diet can be supplemented with a watery gruel of canned or baby food until the kittens are eight weeks old; then they can eat solid food. Otherwise, commercially available powders, such as Kitten Milk Replacement (KMR) and Just Born, will serve in place of mothers' milk.

Can all the kittens eat from the same bowl?

You can provide a large bowl or individual bowls. Watch to see that the timid kittens aren't pushed away by the chow cats. Invariably, the most dominant or pushy kitten gets the best of the food. Also, keep a washcloth handy to clean off the kittens that leap into their bowl.

Why does my friend's cat have to be petted while he eats?

Contact gives him the inspiration to eat. It triggers the warmth and comfort he experienced in nursing.

Have you ever heard of cats selectively eating certain foods they have a deficiency in?

Yes, it's not unusual for cats to be instinctively or randomly attracted to such foods and to convey the message to their people through their behavior.

My family cat has a passion for ravioli and pizza. Are they bad for him?

Not at all, unless he prefers such cuisine to the exclusion of his own food. His figure will let you know if he is being overindulged.

Speaking of overindulgence, what is the best way to control a cat's weight?

Excess weight can stress a cat's heart and cause other physical problems. A little extra flesh can be useful in an emergency, but too much is a definite health hazard. Each cat's "normal" weight depends upon his individual frame and bone structure. But if your altered male cat's tummy swings low to the ground, his sides bulge, and he has lost all his pep, you can be sure he is a "chubby." If your spayed female has a profile closer to a pear than a panther,

she is part of the club. Here is what you can do to help your Chubbette get slim:

1. If your cat is medium-sized at his normal weight (nine to eleven pounds), divide four and a half ounces of food into three daily feedings. Three feedings will make him feel he is eating more. There are high-fiber cat foods that are weight-loss friendly. Sometimes kitten canned food, which is high in protein, is an option. You can also use an organic diet.
2. Don't sneak him any food on the side. Remember, food has calories, and calories add to the body weight; they never subtract! However, if you must give him a snack because you just can't bear his woeful eyes, try some melon—a low calorie snack. But give him only a few slices. Don't overdo it.
3. Encourage him to move around! Play with him, and try feeding him in high places so he has to exert energy to reach his food. You might play with him just before his second or third daily feeding, so he knows that food follows fun.
4. If he is your only cat, get him a friend to help him work out his energy.
5. If you have two cats, and Chubbette devours all of his companion's food, feed his companion separately. If his companion refuses to eat alone, stay with him and stroke him; he may desire contact while he eats. If that doesn't succeed, feed the companion a food Chubbette doesn't like.

Remember, don't cheat and try to sneak in a handful of dry food. Crunchies are especially high in calories, much like cookies. And, last but not least, make sure you are giving him enough attention and he is not eating out of frustration. Double the attention! If you can't spend enough time with him, arrange to have a neighbor or a neighbor's child pay him daily petting-and-playing visits. Love and attention can outweigh his need for food, but you must be sincere or he will feel the difference!

What if my cat is too skinny and not much interested in eating?

If your altered cat is too slim even for a *Vogue* model, first have the vet check him out. Have a stool sample tested for parasites. Sometimes the test has to be repeated a few times for accurate results. It's not unusual for a cat to be treated with medication for symptoms of parasites if there is a strong suspicion of their presence, even if the stool sample is negative.

If nothing is amiss but your cat's appetite, and he's picky, picky, picky, here are some pointers to help fill out his lean, little bod:

1. Try giving some cooked meat or chicken. He may appreciate your personal touch.
2. Try to hand-feed him; then he knows he is really special.
3. Give him several small daily feedings if he is a nibbler. If your other cat objects, feed him more often, too, but don't increase his total amount of food.
4. If your other cat tries to eat Twiggy's food, distract him or feed Twiggy some food his companion won't eat.
5. Supplement his diet with high-calorie vitamins. Consult your vet; he may even carry a supply. High-calorie foods are also available.

If all fails, perhaps your vet can prescribe a medication to stimulate your cat's appetite. If he is your only cat, he may need a companion. The competition and exercise may tickle his tummy.

Make sure that you are giving your thin cat enough attention. He may have lost his appetite because he feels neglected. Arrange for someone to spend time with him if your schedule is too demanding. He may need extra loving care to make him feel special and to stimulate his appetite.

If your cat needs a special diet—for example, for diarrhea or constipation (see Chapter 8, Health; and Chapter 16, Senior Cats)—he may also need some vitamin and mineral supplements. Consult your vet. If your cat has a particular vitamin or mineral deficiency, a well-balanced diet is not enough.

In the meantime, try serving up the following healthy homemade dish. If cystitis is a problem, this low-ash treat is easy on your cat's bladder as well.

Healthy Meow Chow (high protein, high fat, high B vitamins)

3 cups rolled oats
1 cup wheat germ
1 cup bran
¼ cup desiccated liver powder
⅛ cup bonemeal powder
½ cup debittered brewer's yeast or yeast plus
1 teaspoon kelp
¼ cup liquid soybean oil
¼ cup melted shortening or suet
¼ cup flaxseed oil
½ cup powdered chicken broth
ground flax seeds (optional)

1. Mix dry ingredients well (may include some ground flax seeds in place of oats, wheat germ, bran). Toss with oil and melted shortening. Add water or milk (enough to sufficiently moisten and hold together). Sprinkle in some catnip if your cat's a devotee.
2. Toast on lightly greased baking sheets for 20 minutes at 350 degrees. Turn often to brown on all sides.

Store in airtight containers in refrigerator or freeze half of it.

Your awareness of your cat's diet will provide him with a healthier and happier life, and you will reap the dividends together!

chapter eight

taking care of your cat's health

"But my cat never goes outdoors! How could he possibly catch distemper?" Time and time again, clients repeat this question. It is indicative of a cat myth that can have lethal effects on your pet.

A cat does not have to go outdoors to be exposed to feline distemper, otherwise known as panleukopenia, or cat fever. Often an outdoor cat has more immunity to the disease because he is constantly exposed to it. Feline distemper is an airborne virus, and you can be the carrier. It can be transferred to your cat even through your shoes. You never know when you may come in contact with another cat person or object that has been exposed to the virus.

The virus attacks the cat's digestive tract and bone marrow. Common symptoms are loss of appetite, withdrawal, dehydration, lethargy, diarrhea, and vomiting. The disease can be treated, and there are those cats that survive, but it is more often deadly. Why take a chance with your cat's life? Make sure your kitten is vaccinated and check with your veterinarian as to when a booster shot is advisable.

When can a kitten be vaccinated against distemper?

He needs a series of vaccinations that can begin as early as six weeks and extend until he's thirteen to sixteen weeks old. Since a kitten's immunity against distemper forms somewhere between those ages, it is necessary to keep challenging the virus to be sure the immunity forms.

So your kitten's amount of vaccination depends upon his age. If you adopt a six-week-old kitten, you vaccinate him then and again at eight, twelve, and sixteen weeks. However, if you adopt a fourteen-week-old kitten, a vaccination is needed then and again at sixteen weeks.

Suppose I adopt a ten-week-old cat that has never been vaccinated?

Have your vet examine him to be sure he is healthy, and, if so, start the vaccinations.

Can an older cat catch distemper?

Age is not the determining factor. Any unvaccinated cat is a perfect target for the disease. An annual vaccination increases a cat's immunity. The only cat who doesn't need a distemper vaccination is one who has had distemper.

Are any other vaccinations necessary?

Yes. Your cat should be vaccinated against the respiratory viruses. Sometimes these vaccines can be combined with the distemper vaccine. Be sure to consult your vet.

Isn't there a rabies vaccination?

If your cat is going to be outdoors where he will be exposed to rodents or wild animals, a rabies vaccination is mandated.

Why doesn't every cat need a rabies vaccination?

Because in order for a cat to be infected with rabies, he must be bitten by a rabid animal (such as a squirrel, rat, or bat).

Can't a person contact rabies from a cat?

Only if bitten by a rabid cat.

How can I tell if my kitten has worms?

Symptoms can include vomiting, diarrhea, ravenous appetite, runny eyes, dull coat, and a potbelly. Even if your kitten shows no such symptoms, his stool should be tested for worms each time he is vaccinated.

How often should you check a cat for worms?

An annual stool-sample check is sufficient. However, if your cat goes outdoors and comes in contact with other cats, or acquires a new litter box companion, a sample should be tested more often.

Aren't there many different kinds of worms?

Yes, there are several varieties, including roundworms, hookworms, tapeworms, ringworms, and protozoa. People rarely see the adults of roundworms and hookworms, unless a severe infestation causes the cat to vomit or pass an adult worm in the stool. Most often, the eggs must be identified through a microscopic examination of the stool.

How can I tell if my cat has ringworm?

It usually shows up as circular areas of hair loss where the skin appears crusty, and there may be new hairs growing in the center of the ring.

What is the treatment?

It is usually a combination of an oral medication and a topical salve for a cat; for a person, the treatment is usually salve because a person has less hair!

What do roundworms and hookworms look like?

They resemble spaghetti.

Can a person spot any other kinds of worm?

Tapeworm is one that is commonly spotted by people. The worm resembles rice and often sticks to a cat's rectum. The segments are often seen moving, expanding, and contracting. They are rarely identified in a laboratory examination because the eggs don't float in the solution.

What causes tapeworm?

Tapeworms are transmitted from cat to cat with the flea as an intermediate host. The white segments contain many individual eggs. The flea ingests the eggs, which then hatch into larvae and develop inside the flea. When a cat swallows a flea, the flea is killed, but the tapeworm larvae attach to the intestinal wall, grow to adults, and, voilà, more tapeworms.

How are other worms spread?

The eggs come out in the cat's stool and hatch into larvae without an intermediate host; another cat ingests the larvae, which mature in the body. Kittens frequently develop the same parasites as their mother.

What's the treatment for worms?

It's usually in pill form. The dosage and time involved usually depend upon a cat's weight.

How can I tell if my cat has fleas?

Sudden, furious scratching with resulting patches of thin fur or irritated skin is often a giveaway. Tiny black flecks on your cat's fur or on your furniture are also telltale signs. Fleas are rampant, especially in beach areas.

What are the precautions and treatment?

Check with your vet for the best prevention. It's usually in the form of a pill, but sometimes liquid, that can be given monthly. A flea bomb may be necessary if your place becomes infested with fleas. Fleas usually hang out in carpets, corners, under furniture. They are small, dark brown or black, and scurry or fly when disturbed. (Sometimes homeopathic remedies or Chinese herbs are used.)

What's a protozoan?

A protozoan is a single-celled organism. Coccidiosis is the most common protozoan disease in cats. There are two common forms. The most obvious can cause diarrhea with a bloody gelatinous stool and may or may not also cause respiratory symptoms. There is a subacute or chronic form, which is much more difficult to diagnose—it is sometimes necessary to test several stool samples before it is discovered. (See Chapter 9, Litter Box Problems, for symptoms.)

Does a cat need a dentist?

Your cat's teeth should be checked on his annual visit to the vet. Some cats never need dentistry, but teeth and gum problems in cats are not uncommon. There are a few veterinarians who specialize in feline dentistry, but any vet can take care of your cat's teeth. A cat will communicate a mouth or tooth problem by frequently sticking out his tongue, pawing at his mouth, chewing his food with difficulty, vomiting, and/or rejecting his food.

What can I do to keep my cat's teeth and gums healthy?

Cooked chicken necks and backs and dry food (chow) provide good exercise for the teeth and gums. There are also many cat foods available for this purpose. Some cats will allow you to brush their teeth with a special brush and tooth paste. Consult your vet.

I heard somewhere that a cat's anal glands can cause him trouble. Can they, and where are they?

Yes, they can. There is a gland located on either side of the tail at roughly four and eight o'clock. When they fill up and aren't released, itching occurs, and the cat becomes uncomfortable.

What happens then?

The cat tries to empty them out by licking at them (which may cause his throat to become inflamed) or by dragging his rear along the floor.

And if he doesn't succeed?

Then he will be uncomfortable until you take him to the vet, who will squeeze the glands and empty them out. Your cat may need an injection if the glands are severely impacted.

Should I worry if my cat misses a meal?

One meal—not to worry. It may be the weather, or he is being trendy and fasting. However, a few meals or three days without food may be a sign of illness and requires a visit to the vet.

What about vomiting?

Occasional vomiting may result when your cat has wolfed down his food or when he has to rid himself of a fur ball. Regular combing or brushing will decrease the amount of fur your cat swallows when he grooms himself or his companion. Special foods that treat hair balls can contribute to vomiting. Continued vomiting may be an indication that your cat needs medical attention.

How do I keep my cat regular?

Add a bit of butter or Vaseline to your cat's food twice a week to keep him from getting constipated. A commercial gel or powder can be mixed in with his food. Some cats will lap the gel right from the tube.

Suppose my cat's constipation is chronic?

This is a problem to discuss with your vet. It usually occurs with an older cat or one that has sustained a pelvic injury. In some instances a tranquilizer, given in addition to a laxative, will give your cat's muscles the relaxation that he needs to help him move his bowels. Food additives that can help include liver, either raw or cooked, and bran. Special high-fiber cat foods are also available to treat constipation. (See Chapter 16, Senior Cats.)

What about diarrhea?

If it's a fleeting case of diarrhea, mix yogurt or rice in with your cat's food, and don't feed him anything rich. Stick with baby food or cooked chicken. If the diarrhea persists for three or more days, make an appointment with the vet.

Can I trim my cat's nails?

Yes, there's a special clipper you can purchase. A cat's nails should be clipped regularly to prevent hangnails and to protect your furnishings from getting scraped and scratched.

Anything else I need to know about clipping his nails?

Don't clip beyond the red line (the vein), or you'll have a bleeding, hurtful nail and a very unhappy cat. Give your cat a reward, both before the pedicure and after, so he regards it as a happy experience. Be firm but gentle as you press down on your cat's paw pads.

Must I bathe my cat?

Generally, a short-haired cat does not need to be bathed, unless he has been immersed in something very unpleasant and can't deal with it by himself fairly quickly. A long-haired cat may need an occasional bath. (See also Chapter 13, Grooming.)

What about caring for my cat's ears?

If your cat doesn't go outdoors, his ears should remain pink and clean. He can do the minimal grooming and washing that they will need. An outdoor cat's ears are more apt to get dirty. You can clean them out with a Q-Tip moistened in baby oil or lukewarm water.

Can I tell if my cat has ear mites?

If your cat's ears look very dirty and have dark brown deposits, and you find him scratching at them recurrently, your vet should check them out. A microscopic examination is needed to detect mites. Sometimes the debris can be inflammation or plain old dirt.

Anything I should know about his eyes?

If he gets "sleep" in the corners of his eyes, you can wipe them with moistened cotton. Do the same if his eyes are runny. However, for anything remarkable, consult your vet.

Can a cat's breathing be indicative of how he feels?

Yes. When a cat is relaxed or happy, his breathing is generally steady—about eighteen to twenty-two breaths a minute. However, if he is nervous or excited, his breathing becomes rapid, and he may even pant if very excited. This is not a cause for worry, because the cat's breathing goes back to normal when the source of anxiety or excitement is eliminated. But if your cat's breathing is irregular or very fast when there is no apparent source of excitement or anxiety, it may be because he is not feeling well.

Long-term stress can cause your cat's body to become tense where his muscles contract; consequently his breathing is altered. His vulnerability to sickness is then greater, and his stress target will invariably be attacked.

Quite frequently, for example, emotional stress can precipitate an overt asthmatic problem. Rapid breathing and bronchial spasms are triggered by the stress. Or sometimes a cat has an asthmatic or cardiac problem, and the only apparent symptom is bizarre behavior. A medical checkup and chest X-rays often reveal a respiratory problem. Sometimes a sonogram is needed.

A cat may have a chronic urinary or skin problem that doesn't respond to treatment. Frequently the bladder or skin is only the secondary stress target; the primary target is the chest. As soon as the primary problem is treated along with the secondary stress target, the cat will begin to rally.

Generally, with respiratory problems, the underlying factor is stress. It is important that the source of stress be identified and eliminated or reduced. A cat may need a tranquilizer to relieve his anxiety. Paramount, though, is consistent and loving support from his person.

A cat develops poor breathing patterns as a defense for his body against a perceived threat. Whether the stress from the threat is internal or external, the effect of unresolved anxieties often triggers a physical disease. It is not uncommon for a cat to have rapid or bizarre breathing patterns without any related problems or side effects. But if you notice that your cat's breathing is unusually fast or labored without provocation, a visit to the vet is in order.

You haven't mentioned anything about major cat illnesses other than distemper. What about feline leukemia?

There aren't any classic symptoms for feline leukemia. If your cat displays any difficulty in recovering from a minor illness, make an appointment with your vet. A blood sample may be in order to rule out major illnesses, including leukemia. Usually, a feline leukemia test can be done quickly in the office. Feline leukemia can often be treated for an ongoing period, so that the cat can live quite comfortably and happily. The condition can go into remission temporarily or even permanently. Sometimes a cat is hospitalized at first, so treatment can be determined and regulated. Often cortisone is used to make the cat feel more comfortable and to stimulate appetite, but treatment varies. Your cat can be vaccinated against leukemia if he goes outdoors or is exposed to other cats.

Gran-Pa was a six-year-old cat who had feline leukemia. He lived with other infected cats. Since the people also had healthy cats, the leukemic cats were kept separate from the others. Fortunately, the house was large enough to give all the cats enough living space. There was even an outside run for the leukemic cats. Every few months the people would have their cats retested against the disease. What a wonderful surprise when Gran-Pa's results came back! He was negative for leukemia, because the virus had gone into remission. His people were ecstatic. They had worried that Gran-Pa would have to leave all his old friends and make a new start. It would have been like moving to a new neighborhood for him.

Wasn't there a chance of Gran-Pa's spreading the disease to the healthy cats?

No. If the antibody test was negative, he no longer had the virus in his body.

I've heard of cats being carriers of leukemia but not showing any symptoms.

It's possible for a cat to carry the virus without actually being afflicted with it himself. In that case, the cat should be tested every few months to see if the disease has gone into remission.

Should a cat who is a carrier be kept away from other cats?

Absolutely. But if he has been living with a companion, the companion will already have been exposed to the virus.

What about when friends who have healthy cats come over?

For starters, they should leave their cats at home. If they touch the infected cat, they should wash their hands thoroughly before they leave.

Can the infected cat's people visit other people with cats?

Yes, in most instances, but they should also be fastidious about washing their hands. However, they should not visit with people who have cats under stress or whose cats have a low resistance to disease.

How soon can a person get another cat if he loses his cat to leukemia?

He should wait at least three months. Be sure to dispose of old feeding dishes, litter pans, and toys.

Any other major diseases?

FIP (*feline infectious peritonitis*) is an immuno-infectious virus that affects cats. Although it can go into remission, as yet there is no cure. FIV (feline immuno deficiency virus) is another virus that is incurable, but it can go into remission. If bitten, FIV can be transmitted to another cat. Feline epilepsy is a health problem that has become more common lately. Cardiac problems and various types of cancer are also prevalent in cats. With treatment, many of these diseases can be kept under control, and the cat can go on, sometimes indefinitely, comfortable and happy. However, when physical pain outruns treatment and medication, and human contact is not desired, your cat should be put to a peaceful and painless end.

What about cat scratch fever?

Cat scratch fever (from *Bartonella henselae*) is a benign infectious disease that can be carried to a human if the cat has the bacteria. Kittens are usually most likely carriers. There is a test to determine if a cat is infected and medication to treat the bacteria.

What about treatment for the person?

There is medication and symptoms will disappear within two weeks or a month, longer if the person is immuno compromised (consult vet for more information).

Ease Your Cat's Stress after a Marital Split

T. S. Eliot wrote, "A cat is a cat!" And what makes a cat so singular and inimitable is a cat's exquisite sensitivity and splendid psyche. But there are times when these profound and endearing qualities can be a mixed blessing for a cat.

A cat's sensitivity and vast depth of awareness enable him to have a deep emotional bond with his people. So it's not surprising that a cat can be emotionally affected by their behavior. A cat is generally quite vulnerable to major stress or changes experienced by his people. Unless the cat himself decides to initiate change, he is basically a creature of habit and prefers a life of familiar patterns.

Major turmoil and stress are indeed caused by a marital split-up. Although it's usually for the best, it's often a stressful transition. Generally the cat is deeply affected by the change. Your cat's feeling of separation anxiety can be relieved by your awareness of the cat's distress, and you will share in his relief. The following pointers will help to allay your cat's stress:

1. Give your cat support and affection when you're anxious because your feelings affect your cat. Your interactions with your cat will also ease your tension.
2. Try to keep your cats together if they have a close bond. If not, they should live with the person of their choice.
3. If you have a cat on medication, the veterinarian may want to increase the dosage temporarily.
4. A single cat might be comforted by the introduction of a kitten to help ease his loss. (See Chapter 6.)
5. Arrange to have a young child or friend visit your cat if you're out frequently.

6. Activity or exercise that you can do at home to heighten your spirits will similarly affect your cat.
7. It's usually more difficult for an intact cat to cope with stress, so it's wise to have your cat neutered if he's reached sexual maturity.
8. Remember to hug your cat several times a day. Hugging therapy is an effective discharge of pent-up body tension.
9. Another effective way for your cat to release tension is with catnip. You'll notice he becomes peaceful and relaxed after the initial high-energy reaction to the catnip.

People don't have to verbalize their anger or anxiety to communicate their tension and distress to their cats. Cats are very sensitive to body language. When people are upset, tense, or frightened, they carry themselves in an unrelaxed manner. Their bodies reflect these moods, and a cat, through his uncanny ability, can sense this. It's this very ability that enables a cat to survive on the street. A cat's antennae for danger are so very keen. I think of this ability as "Cat-Sense"—a cat is a natural medium for fluctuations in surrounding energy fields, whether human or animal. Consequently, your cat may sense your anxiety long before you acknowledge it.

How to Relax a Cat with Audio Recordings

Because a cat is affected by body language and voice, it is possible to program your input therapeutically when your cat needs extra support.

Audiotapes are an excellent medium for this. You don't have to have a high-tech aptitude. All you really need is a simple tape machine that records, plays back, and shuts off automatically at the end of the tape. A thirty-minute tape is adequate; you can also get a longer one and add to it. With your cat or cats on hand, search out a favorite toy, grooming comb or brush, special security object if any, some soft music, and choice catnip, and refer to the relaxation suggestions listed below.

Relaxation Techniques

1. Set up the tape recorder next to a sunny, comfortable spot where you and your cat can peacefully recline.

2. Choose a time when everyone's biological needs are satisfied
3. If your cat enjoys catnip and it leaves him with a mellow feeling, indulge him in advance.
4. Breathe freely, concentrate on happy incidents, and turn on the recorder.
5. Brush or comb your cat, if he is receptive. If not, stroke him as you talk about his many virtues and frequently repeat his name. Also talk about good times you've all had together and things that make you happy. You might even wish to include some poetry or sing a song if you're musically inclined.
6. When you're ready to sign off, be sure to address your cat frequently by name. The sound of his name makes him feel very important.

For best results play this tape daily. Remember to switch the tape on when you leave. When you go on holiday, instruct your cat-sitter to switch the tape on during and after each visit. One of my clients from Washington, D.C. reports that her cat Raymie settles down by the recorder as his tape plays and that his ears wiggle to the mention of his name. Raymie's self-esteem has grown enormously, and his indiscriminate urination is but a memory.

You may become an audiotape devotee and end up with an extensive library. All the better for your cat, as the repetition of positive evocations will reinforce the support he needs to integrate new coping mechanisms for a healthier life. At some point, you may also use your computer, laptop, or other audio technology to make a CD or other audio recording for use, as audio tape players become harder to find.

Tranquilizers (Anti-Anxiety) or Psychotropic Drugs

When does a cat need a tranquilizer?

When a cat suffers from anxiety that can't be allayed by positive support from his person(s) and companion(s), auxiliary support is needed in the form of a tranquilizer. If the emotional anxiety has triggered a physical problem, it should be treated simultaneously.

Can such tranquilizing medications hurt my cat?

There's no cause for worry if they are prescribed carefully by your veterinarian and given sensibly. There is no set formula; the amount of medication needed varies with each individual cat. It is dependent on his individual stress tolerance and how well his body absorbs the medication.

If one drug does not relieve your cat's anxiety or makes him more hyper, that's your signal to try another. Occasionally a cat has a bad reaction to anti-anxiety medications, but I have encountered only a very small minority.

If you have a negative feeling about sedating your cat, and it makes you nervous and uncomfortable, chances are your feelings will affect your cat. If such is the case, you will have to find another means of relieving your cat's anxiety.

What does the drug do?

When a cat is suffering from continued anxiety, he is hurting inside. If the hurt or discomfort is not allayed, the stress often precipitates physical problems. (Stress targets, such as skin, bladder, heart, etc., vary with each cat.) The tranquilizer relieves the discomfort. If a tranquilizer is used effectively to relieve anxiety, the prognosis is encouraging and optimistic.

Sometimes it is necessary to tranquilize your cat for a short time to meet a particular stressful situation, such as travel, a trip to the vet, or a houseful of guests. In such a situation the tranquilizer would probably be given for a few days at the most. Even for a one-occasion situation it is best to try your cat out on the tranquilizer beforehand to make certain it is effective, or to discover if he'll need another kind. If the vet tranquilizes your cat by injection, it usually works much faster than oral medication.

A cat generally reacts to a tranquilizer in ways that are normal but may disturb you at first if you are not prepared. For example, within forty minutes after the tranquilizer is given, your cat's locomotion may become wobbly and uncoordinated—he may literally stumble about. He may become disoriented and start to talk much more than normally. He may try to resist the reaction of the tranquilizer by running around, but eventually he will settle down and probably go to sleep. A tranquilizer is often an appetite stimulant, so your cat may run to his food dish and demand to be fed, in which case, feed him!

Here is how you can help your cat deal with the tranquilizer: The first time you give your cat a tranquilizer, plan your schedule so you can be at home with him for at least a few hours. Talk to him in a gentle and low voice, and reassure him. Don't laugh at him, because it will increase his disoriented feeling. The more you relax, the easier it will be for your cat. Don't have visitors over the first time you do this: you don't want any added commotion.

Since a tranquilizer relieves internal anxiety, your cat may become more vocal as his anxiety subsides. However, your cat's dosage may have to be increased if he cries a lot. Crying generally indicates he is still too anxious. As your cat becomes more relaxed, he is apt to be more affectionate and desire more contact from you and/or his companion. Also, his appetite may increase, so offer him more food, within reason. Don't add to his anxiety by ignoring his hunger.

Occasionally a tranquilizer causes diarrhea. If so, feed your cat a bland diet—boiled chicken and rice, for example—for a while. If this provides no relief, you may have to decrease his dosage. If diarrhea still persists, a different tranquilizer may be in order. Generally, as your cat's system adapts to the tranquilizer, the above reactions are minimized.

When might the tranquilizer dosage have to be increased?

When your cat is subjected to extra stress—if you go away, if someone stays with you, if there are many loud noises, if you are under extra pressure from your job (your anxiety can affect him), etc.—he will probably need a higher-than-usual dosage to relieve his anxiety and to keep him from having a setback. Sometimes it is hard to realize how much an event affects your cat, because he may have a delayed reaction to the stress. By increasing the dosage, you provide him with extra support. It is worth not risking a setback.

How long will it take my cat to recover? That is, how long will he have to take a tranquilizer?

That cannot be predicted exactly. It varies with each patient. A long-term problem will generally call for extensive treatment—maybe lasting a year or so. It takes time to reinforce new behavior patterns. Don't be upset by a few setbacks! As your cat's stress tolerance increases, recovery from setbacks will be faster, and the setbacks will decrease and finally disappear.

Can my cat become addicted?

No, don't fear that your cat will become addicted. First off, most cats dislike being on any kind of medication. As your cat's stress tolerance increases to where he can interact on a reasonably sustained day-to-day basis, the tranquilizer can be decreased. You will notice when things that previously upset him and triggered his anxiety affect him less and less. If at any time a setback occurs, the tranquilizer should be increased until your cat can cope comfortably again. Only at this time should the tranquilizer be slowly decreased back to the maintenance dosage.

When your cat's catsonality is so well integrated that he can interact and function without incident, the tranquilizer can be stopped. The time element involved—months to years—depends on the extent of your cat's anxiety, your support, and the cat's individual healing ability. When your cat's medication has been lowered to a minute amount and he is frequently snoozing, his stress tolerance has increased to the point where he no longer needs a tranquilizer.

The Wilbourn Way to Pamper a Sick Cat

You know yourself that when you are sick, it is not only the medication that helps you rally, it's how people treat you and how you feel about yourself that raises your spirits and gets you through that horrible cold or flu. True, many an illness has to run its course, but it is so much easier if you have people's care and concern.

Your cat has feelings, too, and his illness will respond to your indulgence and affection. Here are some things you can do to make him feel good.

Home Care for a Sick Cat

1. Talk to your cat a lot, and stroke him whenever you pass by.
2. Comb or brush him if it makes him feel good.
3. Massage his body if he is responsive.
4. Feed him his favorite foods—but check with the vet if there is any question.
5. If his face or body gets smeared with medication, wash it off gently with a washcloth and lukewarm water. Your cat can't feel good if he is all sticky, and when he is sick, he can't groom himself as well as usual.
6. Clean his rear after he goes to the bathroom, if he doesn't have the energy. Use a washcloth or paper towel, and remember to be gentle.
7. If he is a sun worshiper, fix your cat a comfy cushion in the sun.
8. Don't have noisy parties. If you must have visitors, move the cat to a quiet off-limits room.
9. Ask your neighbor to stop in to visit the cat if you have to be out a lot.
10. Hold your cat in your lap and cuddle him a lot if he is willing. Coax him if he is shy.
11. Carry him into bed with you at night.
12. Repeat frequently that you love him and want him to get well. If you can't speak it, think it. He will pick up your positive energy.
13. Play soft, soothing music. It will calm you, and the tranquility will soothe him.

How to Give Your Cat a Pill

The best way to give your cat a pill, if it is not bitter, is to grind it up and mix it into a small morsel of food. If it is bitter, coat it with his favorite gel and then mix it into the food. But if this isn't possible because he won't eat it, try this method:

1. Squat down and back your cat between your legs.
2. Reach your left hand over the top of his head if you're right-handed.
3. Gently grasp him under the cheekbones at the corners of his mouth, with your thumb on the right and your finger on the left. Don't stick your fingers in his mouth and invite an accidental chomp.

4. Holding his jaw thus, rotate his head slowly and gently upward and open. When the lower jaw starts to drop, his head is up enough.
5. Grasp the pill between the thumb and index finger of your right hand, and place your middle finger between the two big teeth of his lower jaw. Gently open his mouth wide and push the pill to the very back of his mouth. Sometimes the use of a plastic device commonly called a "pill plunger" at this point will quickly shoot the pill down. (A pharmacist can also compound the pill into a tasty food paste that will be more palatable, or some prescriptions can be made into a liquid and given transdermally, by syringe, usually in the cat's ear.)
6. With the pill in place, kiss his nose or blow lightly at him as you close his mouth. Such a gesture should fly the pill down. Stroking him under the chin will also cause him to swallow.

With practice, your pill-giving technique will improve until your cat will hardly know you gave him a pill. But if your cat is not a pill taker and the treatment is too stressful for both of you, find out if your vet can provide another kind of medication. You don't want to make your cat sicker when you try to give him a pill!

How to Treat Your Cat Bite

If your anatomy should happen to get in the way of your cat's teeth, and he bites you, here's how to treat your wound:

1. Liberally douse the wound with peroxide.
2. Pack your wound in ice for at least thirty minutes. You can hold the ice against your wound in a towel or plastic Baggie. Ice does wonders to relieve the swelling and stiffness.
3. Remember to keep breathing, even if it hurts, because when your body contracts, the pain becomes more intense.
4. Try to think about something pleasant.
5. Twelve hours later soak your wound in warm water. An application of aloe plant to the area will help to ease the pain and lessen any lingering sore or scab.

6. If there are signs of an infection or a red line working its way across the area near the wound, you should consult your physician.

Don't hold a grudge against your biter. Chances are he warned you (with a swish of the tail, flattened ears, rigid body, hiss, swat of the paw) to keep your distance, but you pushed your luck.

Caring for a Hospital Patient

1. If your cat is hospitalized, arrange to visit him. Think cheerful thoughts while you are there. Even if you can't stay long, he will know you haven't abandoned him. Tell him you will be taking him home soon. He won't understand your words, but he will pick up the positive feeling.
2. If you can't visit him, perhaps you can arrange to have a friend who knows him fill in for you.
3. Take him something from home for familiar smells.
4. Send him postcards signed by his companion and yourself. Of course he won't be able to read them, but the nurses can, and their cheerful feeling will be passed on to him. It is the attention that he needs, and cards are sources of cheer.
5. Bring him his favorite foods, but check them out with your vet.

All that you can do to brighten your sick cat's spirit will aid in bringing him to a speedy recovery!

Your Cat's Postoperative Care

1. Normally, your cat should not be fed for at least two hours after you are home from the hospital. Provide water sparingly the first day.
2. Give your cat a light meal his first day home.
3. Discourage your cat from any climbing or acrobatics. Don't let your cat go outdoors.
4. Contact your vet if your cat appears depressed, lethargic, or unwilling to eat for more than one day after leaving the hospital.
5. If your cat has any sutures, they are generally removed within ten to fourteen days.

6. Distract your cat with food or catnip if he starts to lick away at the sutures. Call your vet if you notice any swelling or increase in tenderness around the suture line.
7. Tomcat symptoms will stop within two weeks if your cat has been altered.
8. If your cat has a companion, take the following precautions when bringing the patient home, to avoid possible hysteria. Remove your cat from his carrier and allow his companion to go inside and smell it out. You might throw in a sprinkle of catnip. The idea is for the companion to pick up the hospital smell in the carrier. The companion will then start to smell like the patient. Otherwise, he could be threatened by the unfamiliar scents and feud with or reject his hospital-released friend. Keep a happy frame of mind and your spirit will prevail!

Daily Care for Cystitis

Cystitis consists of infection and irritation in the urinary tract of the cat; it has a variety of causes. Some examples are viruses, bacteria, and hypersensitive reactions. Stress is the complicating factor and generally precedes an acute attack. The stress can arise from separation anxiety, lack of attention, a change in environment, and many other factors. When a cat is stressed, his body has a tendency to cut down on the total volume of urine processed, increasing the concentration of solid particles and lessening the frequency of urination.

If your cat has shown symptoms of a predisposition to urinary tract problems, here is a basic program of prevention.

1. Eliminate dry food, fish, pork, and horsemeat from his diet.
2. Provide fresh water at all times. Change it at each meal.
3. Canned foods should be beef or chicken because they are of high-quality protein. Ash content should be less than 4 percent. Sprinkle food with salt so your cat will drink more water. There are special commercial foods available, or you can cook for your cat if you follow an appropriate recipe.
4. Avoid canned food with meat or meat by-products, as they may contain pork and horsemeat, which can cause your cat problems.

5. Other recommended fresh foods are egg yolk, cooked steak, veal, beef, and poultry. Kidney, liver, and heart can be fed several times a week. Mix whole wheat or oatmeal cereals, yogurt, and vegetables into your cat's food. Cooked chicken necks and backs provide good exercise for the gums.
6. Prescription acidifiers or vitamin C should be given twice a day to make the urine more acid.
7. Watch your cat's weight. Obesity can complicate his condition.
8. Scoop the litter pan at least twice a day to remove moist clay and stool. Scrub the pan once a week.
9. Your vet will want urine samples to check your cat's progress. To collect a sample, keep a shallow dish by the litter pan, and when your cat urinates, slip the dish under him. Transfer it to a clean container and refrigerate it until you can take it to your vet—this must be within twenty-four hours or the sample will no longer be good. (Consult the vet about a special litter that is used in vet hospitals to collect urine samples.)
10. If your cat responds to catnip, it is a wonderful relaxer. At first he may work out his energy, but soon his body will relax. The more agile your cat's body is and the less it contracts, the happier his bladder will be.

IMPORTANT: Be aware of your cat's daily urinary habits. If there is excessive licking of penis or vulva, straining, spraying, or repeated scratching in the litter pan without a urine flow, contact your vet. Don't delay and hope for the best! Your vet will probably want to check urine samples frequently until it is evident that your cat's condition has stabilized.

Your attentive care will minimize the frequency and severity of urinary tract problems.

chapter nine

litter box problems

I offered my services in a unique way one fall while visiting my cousins in Bennington, Vermont. My cousin had organized an auction to raise funds for the Bennington College Nursery School. Bennington is an artistic community, and the variety of things to be auctioned reflected this spirit. When my cousin asked me if I would offer a different sort of item to be bid upon—a house call to a problem cat—I of course agreed.

There turned out to be a few such cats, their people bid, and my services went to the highest bidder: a couple with a cat named Waffle.

"We are so glad you were included in the auction," the couple exclaimed. "You are exactly what Waffle needs to straighten out his weird toilet behavior." Apparently Waffle refused to take care of his business outdoors like their other cats. Although he urinated outdoors, Waffle chose to defecate on their basement floor, and they were convinced he was doing it to spite them.

The following day I paid my house call and met Waffle and the other two cats. After collecting his case history and observing Waffle's general behavior and presence, I presented my impressions and treatment.

I explained that people commonly but incorrectly conclude that their cat is being a troublemaker when he avoids the litter box, or refuses to "litter" outdoors if given the opportunity. What people generally don't realize is that their cat is deviating from normal behavior in order to communicate. Waffle's bizarre and annoying toilet habits were an attempt to convey a particular message.

Waffle was a tense, altered, two-year-old male who came from a large barnyard litter. Sometimes his eating habits and interaction with his companion kittens had been quite ornery and outrageous; and his general behavior was not as calm and relaxed as that of his companions.

His people mentioned that Waffle had suffered a fractured pelvis through being hit by a car. I told them that Waffle might still be sensitive in the pelvic region. His sensitivity might cause him to prefer the privacy of indoors, where his defecation ritual couldn't be interrupted by any outside source. For his own reasons, he felt safer inside.

So I suggested that, although it was a nuisance to them, it would be best to set up an indoor litter box to replace the basement floor. There was also a strong possibility that Waffle had a parasitical problem that gave him discomfort. I recommended that they have his stool tested by the vet and suggested that an X-ray of his pelvic area might be helpful.

If Waffle felt more secure and comfortable, his personality would become more integrated and he would be easier to live with. Giving him extra attention would help to speed up his progress.

His people admitted that they had thought of keeping a litter box in the basement but decided that it was too much of a bother and that Waffle would have to shape up. It hadn't occurred to them that he might be reacting to his old injury. They told me that they would follow my suggestions and see how Waffle reacted.

Although the vet didn't find any parasites in Waffle's stool, he gave Waffle medication for the most common parasites just to be certain, because sometimes it takes a few stool samples to detect parasites. The vet said that Waffle's old pelvic injury could be a source of discomfort from time to time and would make his litter-box rituals "irritating" to all concerned.

A few months later my cousin mentioned that my auctioned house call was a booming success. Waffle was a happier cat, and his people shared in his happiness to no small extent.

Photo by Joan Johnson

Is it hard to train a cat to use a litter box?

On the contrary, a kitten or cat will instinctively use a litter box, providing you do your part. Use a commercial litter box or plastic dishpan, so the box doesn't fall apart during your cat's session. Be sure to scoop the litter periodically throughout the day.

Why can't I just fill the box up with litter and empty it out once a week?

You could, but the odor might do you in before the week was over. Sprinkling baking soda under the litter helps control the smell, but your cat might object to your sanitation plan and take to "littering" outside the box to show his displeasure, once the box becomes rather dirty. Many cats are quite fastidious about this.

How often should the box be washed, and how?

It is best to wash out the entire box with soap and water once a week.

Litter is messy. Couldn't I use newspaper strips?

Although litter can be messy, newspaper is less absorbent, and your cat may not appreciate damp or gooey feet. Also, paper doesn't cut odor as well as litter.

What's the best litter and scooper?

I prefer the sand or clay litters and also pellets, environmentally friendly without added deodorants, and so do my cats. A plastic or metal scooper does a good job, and you can purchase one at any pet department. There are also orgamic pellet-type litters that are easy to pop back into the box when they land on the floor.

What's your opinion of the closed-in type litter box with a roof over it?

It's fine if your cat accepts the close quarters. Personally, I prefer a kitty "modesty" screen which fits around the box. You can purchase one or even build one yourself—it's pretty easy. (See pages 55–56.)

What about the electric self-cleaning litter box?

Many cats are frightened by it, so not every cat is a candidate.

Is it possible to train a cat to use the toilet?

Sometimes a cat can be taught to use the toilet, if he has the inclination. It is easiest to teach a kitten. Generally it is something that a cat will decide on his own.

Buster is a cat that did just that. He had just moved to a new apartment with his person, Diane, and companions. Before his litter box could be set up, a sudden tinkling was heard from the bathroom. It was Buster! There he was, perched on the toilet seat. He had decided a people toilet would have to do if his wasn't available. He has used the toilet ever since to urinate. If you want to teach your cat to use the toilet, see the directions at the end of this chapter.

How do I discourage my cat from using the tub for his business?

If you have already ruled out the possibility of parasites and know that your cat is not giving you a parasite report, make sure you clean his box frequently, so he doesn't have any sanitation complaints. If your cat still favors the tub, keep a small amount of water in it to discourage him.

Where is it best to keep the box?

I prefer to keep it next to the toilet so the dirty litter can be scooped and flushed right down. Whenever I go to the bathroom, I check the box to see if it needs a scoop. No matter where, be sure the box is in an out-of-the-way spot, to ensure the cat's dignity and prevent interruptions.

Is it necessary to keep a litter box for outdoor cats?

Yes, because it is best to keep them indoors when it gets dark, and darkness shouldn't interfere with nature's call.

Are there times when a cat's urine will smell unusually strong?

That can occur when a cat has a urinary problem or when a cat is reaching sexual maturity. Usually it's the male's urine that is overpowering in smell. (See Chapters 14 and 15, Sex and Breeding.)

What can cause the stool to smell strong?

It can be caused by parasites, or it may be something your cat ate. If in doubt, have your vet check a stool sample.

Is it dangerous for a pregnant woman to come in contact with a litter box?

The only possible danger period is in the first three months of pregnancy, when a woman could be infected with toxoplasmosis. If you are planning to have a baby, have your doctor check your blood for a toxoplasmosis titer (amount of antibody protection). If you have a high titer, you have already been exposed and don't have to worry. If it is negative, or you're skeptical, delegate the litter box responsibilities to your husband.

What about small children?

The only danger would be if they decided to lunch on the litter.

My friend's young kitten refuses to use the litter box. Instead, he leaves piles and puddles around her house. What do you suggest?

First of all, if it is a large house, the kitten may be confused. He should be kept confined in a small area with his litter box and food for a few hours after each meal, until he becomes familiar with where his box is kept and what it is for.

But isn't it unfair to lock him up?

His food will keep him happy for a while. Your friend can sit with him briefly to keep him company, and offer him his toys. He will feel more secure in a small area. A large place can often cause a kitten to become disoriented. Also, the door to where his box is kept should be open at all times.

Could there by any other reason for such messy behavior?

He may be objecting to an untidy litter box, or he could have worms. A visit to the vet is in order.

The vet said my kitten's stool was free of worms, but his stool is very soft and sometimes he defecates in the tub. What is his problem?

If he is a plant eater, that could be your answer. If you don't have plants, there is a possibility he may have a protozoan infection called coccidiosis, which is frequently difficult to detect in one stool sample. You may have to submit samples a few times. If nothing shows up, ask your vet if he will treat your cat symptomatically and start him on medication for coccidiosis. It's not a dangerous medication, and it has no bad side effects.

Giardia is another possibility that should be ruled out. It is sometimes difficult to diagnose and can be contagious to other cats.

My cat loves milk, but it gives him diarrhea. Is there a good substitute?

Try cream or plain yogurt.

What would cause a cat to urinate on the bed?

There are many causes—including a urinary problem, impacted anal glands, and/or approaching sexual maturity.

But why would this cause him to use the bed instead of his box?

If he used the box, how would his person ever know he had a problem? By using the bed, he calls attention to his condition, and his person can't help but notice.

But don't you think he is doing it just for spite because he is mad about something?

I do think something's bothering him. Stress, however, can trigger an emotional problem that can result in a medical problem.

Do you mean something has upset him, which in turn has upset his bladder, and he is indicating this in his behavior by using the bed instead of his box?

Exactly!

What happens if nothing is wrong with his bladder or anal glands, and he has already been neutered?

His bladder may have to be treated symptomatically, and a tranquilizer or anti-anxiety drug may be essential to relieve his anxiety until he feels better. (Refer to Chapter 10, Catsonality Problems.)

My aunt's cat sometimes misses the box when he has a bowel movement. He is an outdoor cat and litters outdoors most of the time. What can be done?

He may need a larger litter box. If that doesn't work, newspaper can be spread around the box to spare the floor, or his present litter box can be placed in a larger one to field his deposit.

Any other ideas?

Sometimes this problem can be the result of an old injury. That can be detected by an X-ray of the pelvis and rear legs for evidence of injury, arthritis, etc. If any inflammation shows up, he can be put on medication.

What will this do?

It will make him more comfortable. If he has had trouble crouching because of discomfort, now he will be able to assume the squatting position, and his aim will be right on target.

My cat sometimes leaves bits of stool in the hallway that leads from the bathroom. What is his problem?

It could be that his stool is hard, and he can't get it out in one try in the box, so as he leaves the bathroom, the added movement causes the stool to drop off. Or if he has lots of fur, the stool could accidentally attach to the fur, and your cat would then shake it off.

So what can I do?

Give your cat a commercial bulk laxative such as Siblin or Metamucil or a gel such as Laxatone. Consult your vet for instructions. You might also add butter or Vaseline to his food a few times a week.

Last week I noticed green leafy segments in my cat's stool. Is this normal?

It is absolutely normal if you have plants and your cat munches on them. If so, hang up your plants and buy him some kitty grass.

Is there anything you can do for a cat who has little or no ability to pass stool?

That problem is commonly called megacolon, and if it exists, you need the advice and assistance of a very sympathetic veterinarian who can show you how to assist your cat manually.

A large percentage of my cases involve cats with abnormal litter box habits. Because cats are usually so fastidious, people become quite exasperated when their cat's behavior deviates from the norm. Many a client tells me that I'm their "last hope." They have tried everything, and they are afraid they will have to put their cat to sleep if I can't come up with "the" solution or an optimistic prognosis.

Doesn't it have a lot to do with the people's efforts?

Absolutely! Sometimes the problem has gone on for so long that they are at their wits' end and will settle only for an instant solution. But even if I know why their cat is not using his litter box, it usually takes a while to get him to resume or initiate normal toilet behavior.

So what you're saying is that the people have to be patient and understanding. Isn't that asking for a lot, especially since they are probably forever cleaning up their cat's messes?

For some people it is asking too much. Others can cope with the situation and realize that their cat's been uncomfortable and soon will feel better and that his toilet habits will reflect this.

What do you suggest if it is primarily an emotional problem and the people don't want to deal with it?

I recommend that the people find a new home for their cat.

Won't the problem continue there?

When the medical problem, if any, is treated, a new and loving environment can put things right, since it will be minus the emotional stress the cat has been vulnerable to in his present home. Not every person and cat are meant to live together, and sometimes both can benefit by a change. The quicker a person can come to grips with this, the easier it is. The longer the problem exists, the longer it takes to correct it. (See Chapter 10, Catsonality Problems.)

If Your Cat Prefers to Share Your Toilet

There are quite a number of indoor cats who prefer using the toilet to using a litter box. These cats reject their box and, on their own initiative, seek out the toilet. However, sometimes they use the toilet for urination and their box for defecation. There are a few variations on this theme.

If your cat has not discovered the toilet but you would like to lead him to it, try my plan of action. First, let me point out that I wouldn't try converting an older cat or an ailing cat to the toilet. It would be easier with an adolescent cat or a kitten. Here is what you do:

1. Take an aluminum-foil pie pan (one big enough to fit), and place it in the opening of your toilet seat. Make sure it is secure and won't fall in under pressure.
2. Fill the pan one-third full of kitty litter.
3. Put your cat's litter box out of sight.
4. Introduce your cat to his new site. As you are holding him, scratch the litter with your fingers and talk to him, so he gets the idea.
5. If he starts to look for his litter box, take him to his new site again, and stroke him gently so he forms a pleasant association with it.
6. After he has used his new site several times successfully, make a small hole in the pan. Several days later, increase the size of the hole. Slowly but surely, increase the opening until he gets into the habit of spreading his legs and body wide enough so his feet are perched on the toilet seat.
7. When you are sure your cat has the idea and habit strongly reinforced, you can remove the pan completely. Now your cat is a toilet user!

If, however, your cat repeatedly uses the floor to show you he prefers not to share your toilet, give in and don't deprive him of his litter box. After all, it is his choice, when it comes right down to it!

chapter ten

catsonality problems

"What's your most unforgettable case?" is a question I hear frequently, especially when I am being interviewed by the media. Because I have seen so many cases in which cats and their people have come through some very difficult and trying times successfully, I have decided to devote this chapter to a few memorable ones.

As mentioned before, when a cat's daily routine is suddenly interrupted, his behavior may be affected. Just how much will depend upon his "catsonality" (a term I coined to describe the sum of an individual cat's characteristics while avoiding anthropomorphizing.)

The important element in a feline catsonality is self-esteem, which develops during kittenhood and is reinforced in every phase of the cat's life. A healthy relationship with the mother cat, positive interactions with littermates, and a stable, loving cat–person relationship are the elements that affect a cat's good feeling about himself and give him a sense of security. If any of these early relationships are missing or negative, he becomes a vulnerable target for ongoing anxiety and stress. It is the cat's reaction to stress that manifests itself in his behavior patterns and affects his catsonality.

I have already discussed how a cat can have an uncomfortable reaction to separation from his people or companions, to new companions, to a change of environment, to the onset of sexual maturity, and to other developments. But now I want to point out how these reactions can trigger acute and long-term problems, especially if a cat's catsonality structure is severely lacking in self-esteem, so that he always feels insecure.

The Anxious Attack Cat

Samantha is a spayed, declawed, nine-year-old cat whose kittenhood was not made in cat heaven. Her person, Julie, bought her in Central Park from two little girls. She was told that Samantha was from a large litter in a small apartment where none of the cats were really wanted.

Although Samantha was always somewhat tense and not very affectionate, she never had any remarkable catsonality problems until she was seven years old. At that time Julie adopted Pansy, an insecure, abused female dog. Samantha tolerated Pansy as long as she kept her distance, but Julie noticed that Samantha now appeared markedly more tense.

Julie contacted me because Samantha had attacked one of her friends. Several months before, Samantha had attacked a neighbor. At the time, Julie had several visitors, and she attributed Samantha's attack to the excitement. However, this time she and her friend were quietly talking when Samantha struck out. Julie was terrified that Samantha was going off the deep end and didn't know what she could do to make her feel better.

The evening I paid my house call to Samantha was not one of my most relaxed visits. Julie had Pansy in the kitchen, the entrance to which was secured by a gate. Pansy voiced her feeling about her captivity with several barking bouts. As I sat on the sofa, Samantha appeared and stared at me from across the room. Her expression and body were as rigid as ice. I concentrated on breathing deeply and thinking calm thoughts so I wouldn't add to Samantha's anxiety. I knew that just my presence was enough to make me suspect.

As I took Samantha's case history, she walked over to look at me. I tossed her a catnip toy I had brought along. She immediately pounced on it and kicked away at it with her hind legs. This intrigued Julie, who mentioned that Samantha was not usually impressed with toys. The toy kept Samantha busy for a while, but she soon turned her attention back to me—the intruder, her present source of anxiety. She decided to join me on the sofa and spread herself out on my clipboard. By the way she lashed her tail, I knew her anxiety was growing. Although she was curious and wanted to get to know me, the information that Julie was imparting made Julie feel anxious, and this feeling was absorbed by Samantha, who was a sponge for any kind of anxiety.

I asked Julie to give Samantha a snack to divert her attention from our conversation, which Julie promptly did and Samantha promptly ate. While I finished taking Samantha's case history, Samantha nibbled away, and then I gave Julie my impressions and recommendations.

I told Julie that Samantha's unstable kittenhood had made a huge dent in her self-esteem, and it was this insecurity that caused her to be a tense and difficult cat. This confused Julie because, she said, she had known several cats who had had a bad start in life but didn't continue to have unbearable problems like Samantha. I explained to Julie that some kittens have an inherent ability to cope with stress better than others; their coping also has much to do with what immediately follows their kittenhood. Julie replied that she had always been good to Samantha and had never knowingly done anything to hurt her.

I told Julie I was sure she had tried her best, but unintentionally she had done things that had made it hard for Samantha. For example, she had had Samantha declawed at several months. Samantha was old enough to be aware of her loss, and this added to her insecurity. A cat with a difficult catsonality invariably becomes more difficult without front claws to defend itself. True, cats can't scratch anyone without front claws, but the bite they may resort to can be a lot more serious than a scratch.

The single-cat syndrome also contributed to Samantha's problems. Julie had tried to give Samantha a kitten when she was younger but had been unaware of the proper introduction technique. Samantha had totally rejected the kitten and became impossible to live with until Julie found the kitten another home.

When Pansy had arrived on the scene two years previously, she had indeed been a threat to Samantha. A dog with low self-esteem only added to her own. Pansy produced another drastic change in Samantha's lifestyle. Any dog

would have been threatening to her, but Pansy wasn't just any dog—she required a great deal of Julie's time and attention because she had been abused in her previous home.

With these changes, and since she was getting less attention from her person, Samantha became more sensitive to anxiety and excitement. It all culminated in the attack on Julie's visiting neighbor—a neighbor who seemed disruptive to Samantha because she was quite talkative and demonstrative, and who was also unfamiliar to her.

Julie said she could now understand Samantha's behavior better, but still couldn't understand why Samantha attacked her friend.

I explained to Julie that Samantha's stress tolerance was very low and that therefore she was quite unpredictable. Sometimes she had a delayed reaction to a past distressing incident. In this particular instance Julie had been away for several days; during that time she had had a friend look after Samantha and Pansy. This filled Samantha with separation anxiety, and she undoubtedly also picked up on Pansy's insecurity about Julie's absence. This pent-up anxiety exploded on Julie's friend. She became the present source of Samantha's anxiety. The friend was at a very uncomfortable stage of her working career and was relating her feelings to Julie. She became the perfect target for Samantha!

Now that Julie had a clearer picture of why Samantha behaved the way she did, she could prepare to give Samantha the support she so desperately needed. I explained that Samantha needed constant praise and attention, but that Julie should never try to stroke or hold Samantha too long, because then she would become overstimulated and might strike out. Julie would be able to feel when Samantha had had enough stroking: either Samantha's body would tense, her tail would swish, and her ears would flatten, or she would retreat.

Praise was important because it would make Samantha feel good. Telling her how pretty she was would set up a positive feeling. The words would make Julie feel good, and she could transfer this feeling to Samantha.

However, this support from Julie would not be enough. Samantha would need auxiliary support from a tranquilizer to relieve her anxiety. Julie was rather skeptical about tranquilizing Samantha, but she realized that she could not be around every minute to help Samantha cope. Nor was Julie free of anxiety herself.

Several days later Julie called with a progress report. It had taken a few tries to find the comfortable dosage of the tranquilizer for Samantha, but Julie reported that from then on Samantha had become a changed cat.

Suddenly she was interested in things around her. She was more relaxed with Pansy; and, for Julie, the best part was that Samantha was more affectionate and almost cuddly. I told Julie that Samantha was off to a grand start but that it would take a while before her behavior was adequately integrated.

There could be setbacks. If there were, Samantha's dosage would have to be increased for a few days before returning to maintenance. Then, as her stress tolerance increased, Julie could slowly taper the dosage, but it would probably take five or six months before Samantha could comfortably cope without a tranquilizer.

A month passed. Julie reported to me that Samantha now exhibited wonderfully friendly behavior. The neighbor whom she had once attacked paid a visit, and Samantha behaved like a princess. She climbed into the neighbor's lap and showed no resistance to the neighbor's attention.

This was a major victory. Julie also mentioned that Samantha had previously never been able to have a daily bowel movement, but now she was regular to the day.

I explained to Julie that now that Samantha was more relaxed—both emotionally and physically—it was easier for her to have a bowel movement because her body wasn't as tense and contracted. Things were going very well for Samantha, alias Attack Cat, and would continue as long as Julie combined her support with the aid of the tranquilizer.

When Samantha was well enough not to need a tranquilizer, her self-esteem would be at a point where she could cope with temporary separation anxiety and other changes in her life. She might need the occasional tranquilizer to tide her through an anxious situation, but it would be a variable instead of a constant.

Changes Can Cause Great Anxiety

A kitten who isn't the pick of the litter may experience some hard feelings, but the lone kitten who remains after all his littermates and friends are gone is usually already a victim of low self-esteem. Phaedra was such a kitten. He spent a week in a cage in a pet shop as the solitary kitten before his people appeared. Phaedra was three months old then and became the only cat in the household. For three months everything went well, until his people decided he needed a companion.

The new companion, Nico, was a three-month-old Persian. Nico was exceptionally outgoing, and his people found it hard to keep from praising and petting him. The more Nico shone, the more Phaedra withdrew; his appetite fell off, and his coat lost its shine. Soon after, Phaedra's people split up, and his behavior was even more noticeably affected. He began to knock things over and to scratch at the stereo speakers and walls. When Pat, his person, called me, she was beside herself with worry.

I made a house call the next day. From Pat's preliminary description of Phaedra's behavior, and the cat's facial expression and body posture, I could tell that Phaedra was a very sensitive cat and that it would be a while before he felt better about himself. I told Pat that Nico was the present thorn in

Phaedra's side. His experience at the pet shop had left Phaedra low on self-esteem to start with, and when Pat's husband departed a few weeks after Nico appeared, Phaedra transferred his separation anxiety to Nico. Although Pat now had a roommate whom Phaedra liked, he still hadn't recovered from losing his male person.

I told Pat that she should devote a lot of time to drawing Phaedra out. She would do best to refer to Nico as "Phaedra's friend" and provide Phaedra with catnip treats and tasty snacks to help make him feel very special. Nico was reaching sexual maturity, and I explained that this was also a source of anxiety to Phaedra. Pat should make an appointment to have Nico altered very soon so that Phaedra's stress tolerance wasn't pushed to the limit.

When I asked Pat about Phaedra's health, she mentioned that he occasionally had had trouble with his gums and that lately this problem had recurred. I explained to Pat that Phaedra's present emotional stress also affected his most vulnerable target—his gums—so she should ask the vet to attend to this problem. As soon as she gave Phaedra the support he so much needed, he would begin to feel better both emotionally and physically.

My next conversation with Pat was optimistic. Her special efforts to make Phaedra feel better had paid off. She had had Nico altered, and already Phaedra's spirits were better. I told Pat that she appeared to have things under control but to contact me if there were any setbacks in Phaedra's behavior.

Six months later I had an urgent telephone call from Pat. She had moved to a new duplex apartment, and Phaedra had become impossible to live with. Every night at bedtime as soon as Pat turned out the light, he prowled the first floor, howling and knocking things over as he tried to climb the walls.

I scheduled an appointment for the following evening. Pat's new apartment was an older duplex off a small courtyard, quite different from her previous modern apartment building. She had her bedroom on the second level and her living area below. Although the living room had a huge skylight, the new apartment was not as sunny as her other apartment.

Pat told me how she tried to keep Phaedra with her in the bedroom at night, but there wasn't any door; and even if there had been, he would probably have tried to scratch it down so that he could go downstairs. As we talked, Phaedra and Nico both made an appearance, but it was Phaedra who appeared the most curious. Pat had mentioned that she often heard animals, which she thought were other cats, run across the skylight, which made this apartment noisier than the other. I commented that Phaedra might be reacting to the fleeting cats. His nocturnal prowling might be an attempt to reach the intruders.

In the midst of my explanation we heard a scurry across the roof, and Phaedra, who had disappeared upstairs, appeared in a flash, jumped to the mantel, and tried to scale the wall. It was as if he were acting out my very words. Pat was amazed because, besides herself, I was the first to see Phaedra in action. Perhaps on my behalf, the bedtime intruder had decided to make an early run. But what was most remarkable was that Phaedra's reaction was a clear example of cause and effect.

Now that there was no doubt about why Phaedra behaved in such a hysterical manner, something had to be done to alleviate the problem. Pat wasn't in a position to move again, and she could hardly control the traffic across the roof.

I recommended that Pat play soft music in her bedroom each evening for a calming effect. Phaedra would have to be started on a tranquilizer to relieve his anxiety, and as his stress tolerance increased, the rooftop runners would not be so threatening to him. I also recommended that she leave the light on in the living room, because it might help to keep the intruders away. Last but not least, Pat should give Phaedra a lot of positive support. Frequent reinforcement would allay much of his hysteria. Pat agreed to give it all she could—anything for a decent night's sleep.

Pat's efforts were not in vain. Phaedra responded to the treatment and soon spent his bedtime hours on Pat's bed instead of running frantically about downstairs. Because he was relaxed, he became more affectionate with both her and Nico. Pat was very pleased, as she was rightfully afraid that Phaedra's anxiety could have been displaced to Nico. I told Pat that she would have to keep Phaedra on a tranquilizer for a while, along with continuing her positive support.

Phaedra had definitely had his run of traumatic experiences. When he had just managed to develop a positive interaction with Nico, his whole world was moved to a very different apartment, which presented a huge threat. When his territory was invaded by the rooftop intruders, what was left of his presence of mind, and self-esteem, quickly disintegrated. But now Phaedra was on his way. In time, he would be strong enough to cope without the aid of a tranquilizer and, who knew, maybe by that time the rooftop would be closed to runners!

There's Usually Anxiety Behind Unusual Behavior

Abby was a two-year-old spayed female who lived with her person and a companion/daughter cat, Tiger. Abby adopted her person. She made her entrance through an open window and a few days later delivered a litter of kittens. Before she could be spayed, she gave birth to another litter. Her person was able to place all the kittens, fortunately, but decided to keep Tiger so Abby wouldn't be alone.

Abby's person, Muriel, had read my books and wrote to me in hope that I could shed some light on Abby's problem. Most of the time Abby acted like any other cat, but suddenly, for no apparent reason, she would take off like a rocket, screech to a halt, and lick furiously at her legs and paws. The expression on her face and the tension in her body was enough to let Muriel know that Abby was in discomfort. Muriel was afraid that if these attacks became more severe, Abby might go into convulsions. Her vet had diagnosed Abby's condition as hyperesthesia (increased irritability of the skin to any external stimulus), tried her on steroids, which worked for a short period, and recommended a tranquilizer. However, Muriel was reluctant to try a tranquilizer.

I telephoned Muriel because they lived in Wisconsin and there was no way for me to make a house call. When I asked my case-history questions, Muriel told me that part of Abby's diet consisted of raw smelt and liver, which she adored. I explained that fish and liver can sometimes contribute to a nervous condition and that she should eliminate them from Abby's diet, substituting cooked meats and poultry. Muriel mentioned that she gave Abby brewer's yeast, which she felt was good for her. I agreed. brewer's yeast is high in B vitamins, which help to alleviate stress. Muriel said she would give it to Abby on a regular basis and follow my diet recommendations. I emphasized that she should give Abby a lot of contact and compliments, and try to keep her away from loud noises and excitement.

Abby probably still had leftover anxiety from her street experiences, and whenever she became anxious, these memories would be triggered and cause her to display her bizarre symptoms. With the proper diet and emotional support, Abby's self-esteem would increase and her physical symptoms would slowly disappear. Also, as her vet recommended, a tranquilizer might be the answer, but we would keep it as the ace in the hole.

Muriel's next letter was a pleasure. Although at first Abby missed her favorite foods, she soon adapted to her new diet regimen. Muriel doubled the attention she gave Abby; she made sure she touched or spoke to Abby whenever she passed her. Muriel was afraid Abby would have a bad reaction when some drilling was done in the basement, so she fixed Abby up a cozy hideaway in the bedroom closet. What a delight it was when Abby made it through the ordeal without any problems.

Three weeks had passed without one setback. This was a record for Abby! Although Muriel realized that Abby was still vulnerable, her stress tolerance had increased and her bizarre behavior was on the decline.

Bizarre, Destructive Behavior—What's Wrong?

A cat's person often becomes very perturbed when his cat's behavior turns bizarre and destructive. Roxy was a cat who turned out to have a physical problem that caused her to deviate from her regular toilet habits when she was stressed. However, her person, Bruce, thought at first that her behavior might be motivated by spite. He had adopted Roxy from a shelter when she was fourteen months old. The shelter had spayed her and told Bruce that she had lived with a family for eleven months before her arrival at the shelter.

Soon after Bruce took Roxy home, Roxy began to urinate on his bed. If he closed his bedroom door, she would urinate on one of his two roommates' beds or on the carpeting in the living room. Roxy generally "misbehaved" when she was alone. Bruce had her checked by the vet, but her urinalysis and other tests showed no signs of a medical problem.

When I made my house call, I was quite taken with Roxy, who was both playful and affectionate. Bruce told me how much he adored her, but he couldn't go on living with her indiscriminate puddles. If she had a physical problem, he wanted to take care of it, but if not, how could he change her destructive habit?

Roxy did have a physical problem. A thorough medical examination showed that she had severely impacted anal glands. When she was stressed, the problem flared up, causing her discomfort, and she urinated outside her box to call attention to her problem. Roxy's anal glands were her stress target, and her separation from her former people and long stay at the shelter had made her feel quite insecure, a vulnerable target for stress. Quite likely she had gone through many heats before she was spayed, and the heats had activated the problem with her anal glands.

Bruce wondered why Roxy's problem hadn't been discovered at her previous exam. I told him that the former vet may simply not have thought to check out her anal glands, as many cats display somewhat different symptoms with blocked anal glands—for example, dragging their back ends across the floor or rug.

Roxy's anal glands were emptied out and she was put on medication to stop their inflamation. I recommended that Bruce and his roommates keep their bedroom doors closed while they were away from home and roll up the living-room rug until Roxy's condition had cleared up. She would probably need a few more checkups to get her problem under control. I suggested that a companion might cure her loneliness. But Bruce should make sure that Roxy's anal glands were healthy before he considered a new addition. In the meantime he should try to devote extra time to Roxy, so she would feel happier and healthier.

Roxy had a couple of slight incidents after that, and, indeed, it took a few more checkups before her problem subsided. I told Bruce that when Roxy was under any unusual emotional stress, her anal glands were a prime target, so he should try to anticipate such situations and have Roxy checked out regularly.

Super Separation Anxiety

Separation anxiety, as well as an extreme overweight condition, was at the root of another cat's avoidance of her litter box. Pippilini was a spayed

four-year-old cat who lived with her people and companion cat, Bollitas. Pip had always had a weight problem, but when her people separated for several months, she really packed the food in. At first she appeared happier when her people became reconciled and they were all together again in their former home.

But soon Pip started to leave puddles and piles on their Oriental rugs. Pip's people went along with her behavior for a while, but soon their rugs were at the cleaners' more often than on their floors. The evening I paid my house call, her male person, Bernard, greeted me at the door with one of Pip's piles in a paper towel. She had just deposited it on the foyer rug.

During my visit Pip made a few appearances, and I was quite flabbergasted at her size. What a natural for Weight Watchers! Bollitas, her companion, was a large cat but not a fat one. They had a compatible relationship, but he was the dominant one. He usually had the choice of spots, and it was Pip who groomed him.

I finished my case history and gave Pip's people, Bernard and Vicky, my impressions and recommendations for treatment. They agreed with me that Pip's kittenhood had not been stable. They had adopted her from a person with an apartment full of cats. It took Pip a while to relax and settle into her new home. Fortunately, she and Bollitas, who was the first cat, accepted each other almost immediately. However, her self-esteem was already on the minus side because of her former environment.

Pip's overeating was a symptom of her feeling of emptiness inside. When Bernard and Vicky separated, her anxiety plunged her into food even more. The family reunion was a plus, but it meant added stress and hysteria for Pip, since there were many house parties and houseguests. Much of this entertaining was business oriented and had to be done, but Pip's anxiety level was pushed to its limit. She experienced a spasm in her bladder and rectum and avoided her litter box to draw attention to her discomfort.

Bernard and Vicky couldn't really change their lifestyle much for the present, but they could treat Pip's anxiety and enable her to cope with the high energy level of visiting guests. I recommended that, most important, they give Pip extra attention and compliments and that they also start her on a temporary tranquilization program. The happier Pip felt, the more integrated her behavior would be. At my suggestion they had already had Pip examined at The Cat Practice and had started her on a diet to reduce her weight but not make her feel deprived. I told them that it might be best to divide Pip's daily food into three or four meals so that she would feel that she was eating more instead of less. Although their continuing positive support would increase Pip's self-esteem, I explained that she would need the tranquilizer as an auxiliary to relieve her anxiety. This would give her stress tolerance the added boost it needed to cope with the bustling visitors.

As Pip gradually relaxed, her spasms would cease, and she would return to her litter box on a full-time basis. In order to relax, Pip would need a tranquilizer regularly for several months, but as her catsonality showed signs of improvement, her dosage could be slowly tapered and finally stopped. Bernard and Vicky were somewhat puzzled about the tranquilizer, but they agreed to give it a try.

Vicky called me during the next week with an excellent progress report. Pip was playful; she spent more time with Bollitas and even appeared more assertive. But, best of all, she didn't have any carpet incidents. Vicky wondered if it was safe to put down their freshly cleaned carpets. I told her to wait until Pip was more stable, as she was apt to have a few setbacks in the beginning.

Pip's next report, a month later, was another gem! Now, when they had houseguests, Pip was able to voice her opinion with a hiss instead of internalizing it into a spasm and leaving her message on the rug.

The Tomcat Syndrome

Quite frequently the tomcat syndrome causes a male cat to exhibit anxious and/or hostile behavior when he feels threatened.

Cary Grant was my first dramatic tomcat-syndrome case. A street orphan, Cary had appeared at The Cat Practice when he was about fifteen months old. Apparently he had been abandoned just at the point when he reached sexual maturity. His subsequent traumas—the stress of street survival, abuse, tomcat brawls—caused him to be quite the Siamese warrior. Cary's stress tolerance could not cope with anything that challenged his tottering peace of mind. When at all threatened, he would attack the source of his anxiety to wipe out the uneasy feeling it gave him. His means of attack was to lunge and sink in his teeth.

Cary's targets were anxious and high-energy people, primarily men, as energy usually emanates more strongly from males than from females. Unfortunately, my veterinarian husband, Paul, fit into that category and was one of Cary's major thorns. The sound of Paul's voice could wake Cary from a deep sleep, and he would scream with menacing rage.

Cary occupied our office. Consequently, it became Cary's office. Spurred by necessity, Paul then built Cary his own private room in a sunny, secluded part of the hospital. Treatment consisted of constant positive support (Cary had a nurse sitting with him practically every waking hour), Clark Gable (his adopted kitten), and a carefully monitored tranquilizer schedule.

Cary's prognosis was dim. But sixteen months later he had become a happy, mellow cat and was off medication; and he and Clark Gable were adopted by Maria, Cary's first and favorite cat nurse.

My late cat, Sunny-Blue, was another prime example of a full-fledged tomcat whose sexual stress, street traumas, and abandonment syndrome left him low on self-esteem and high on aggressive, hostile behavior. (See Chapter 15, Sex and Breeding: His, for the history of Sunny-Blue.)

Like Cary, Sunny-Blue completely recovered from his tomcat-syndrome days. But because of his particular catsonality, he did have extraordinary quirks. His need for attention became overwhelming. He practically tackled us if he felt he was being ignored and/or yelled nonstop until one of us picked him up and hugged him.

Sunny-Blue was so exquisitely sensitive to touch, sight, and sound that at the drop of an object he would appear in a flash to investigate or bat the object around. He darted through any open door, including that of the elevator and the neighbor's apartment. His curiosity had no limits!

Although he eventually achieved decent self-esteem, it still took a while for Sunny to feel that Baggins and Paul and I were not going to abandon him. We had to continue giving him the contact he demanded and needed. Fortunately Baggins was twice the size of Sunny. When he felt Sunny was out of line, he wrestled him down. This workout relieved Sunny's pent-up energy and allowed Baggins to go off for his snooze in the sun.

Occasionally Sunny became anxious if our energy level was high, but instead of attacking, he stared at us and gave a few howls to let us know he was upset. I knew his self-esteem would blossom and that he would become a very self-possessed Sunny-Blue.

A Different Story

J.J. was a six-year-old altered male cat whose self-esteem was affected by stress from the tomcat syndrome and multiple other factors. J.J. was three months old when our friend Dick adopted him from a garbage can. Dick was away most of the day, and his neighbor frequently looked in on J.J. and brought him treats. When Dick realized he could not give J.J. the proper attention, he gave J.J. to his neighbor.

Unfortunately, a year later J.J.'s person became sick and had to be committed to a mental hospital. J.J. moved in again with Dick, who now lived with his new wife, Linda, and her young, spayed female cat, Chin-Chin. The two cats took to each other. But J.J. was still a tomcat, and Chin-Chin couldn't deal with his bites on her neck and the rest of the tomcat ritual. J.J. was neutered shortly thereafter, and the family moved to another and larger apartment. J.J. was affectionate with Dick but wary of visitors.

The new apartment had extensive carpeting, and within a few months Linda noticed it smelled of urine in various spots. One day she caught J.J. "in action" on the living-room carpet. She and Dick decided that perhaps J.J. was signaling a need for more attention. They made an effort to spend more time with him, and for a while the carpet was untouched, but within a month problems began again.

When Linda told me about J.J.'s habit, I told her that because of his past psychological trauma—abandonment syndrome, separation anxiety, and overextended tomcat status—J.J.'s self-esteem was low. Whenever he became anxious, his unresolved feelings were triggered, he got a spasm in his bladder, and he urinated on the carpet to draw attention to his discomfort. I suggested that he be started on medication to ease his bladder, that they

continue to give him extra attention, and that he have a tranquilizer to allay his anxiety.

Dick and Linda were reluctant to tranquilize J.J., but they did have him examined by the vet. Although J.J. had no clinical problems with his bladder, the vet treated him symptomatically and prescribed medication for the bladder. Once again J.J.'s behavior improved, but not for long. When Chin-Chin died of a fatal kidney disorder, the stress was again too much for J.J. He missed his good friend, and his feelings of separation anxiety triggered subsequent upsets with his bladder. Consequently he registered his anxiety on the carpets.

Dick and Linda understood his sorrow and adopted a young male kitten, whom they named Felix, to cheer J.J. up. J.J. accepted Felix immediately but still had occasional accidents.

Linda called me one afternoon with the good news that she was pregnant. She and Dick were ecstatic, but she realized that J.J. might have an ambivalent reaction to her pregnancy and to the baby, and his bladder could be in constant turmoil. Linda and Dick decided that they would give tranquilizer a try, to see if that was what J.J. needed. With extra positive support, and a tranquilizer, J.J. became a changed cat. Not only did he confine his business to the litter box, but his body relaxed and his facial expression became softer. Dick and Linda were tickled with J.J.'s new catsonality and especially pleased that J.J. didn't christen their freshly cleaned carpets. I reminded them that J.J. would have to be on a tranquilizer for several months, and they would probably have to increase his dosage when the baby arrived.

Six months later J.J. had a new baby person named Scott. J.J.'s behavior was remarkable! At first he checked Scott out from a safe distance, but soon he hovered close by. He especially seemed to take vicarious pleasure in Linda's nursing of Scott.

Shortly after Scott's arrival, Chow-Hound, a four-year-old female dog, joined the family circle. J.J. was skeptical about her, and I recommended that his tranquilizer be increased for a while. Fortunately, Chow-Hound's temperament was mellow and didn't challenge J.J.'s stress tolerance.

One evening we paid a social visit to Dick and Linda. When J.J. curled up next to me on the sofa, I could hardly believe he was the same cat I had met two years ago. His relaxed body and calm facial expression told me how greatly his stress tolerance had increased. I was sure that as time went on, J.J. would become more and more a self-possessed cat.

Low self-esteem and insecurity, as these cases illustrate, can severely affect a cat's ability to cope successfully on a daily basis. True, some cats who have had an unfulfilling kittenhood and/or other stressful experiences are not insecure. Others with low self-esteem manage to function without incident if they lead a life of only moderate stress. But most cats are adversely affected by their early unsatisfying experiences, and their self-esteem reflects their past deprivation.

In the cases I have described, the cats, when challenged or threatened, exhibited lingering catsonality problems. They were able to function effectively only when their stress tolerance increased remarkably as a direct result of supportive treatment from their people.

Resolving Feline Aggression

When a cat displays aggressive behavior, it is symptomatic of inner stress. The cat is uncomfortable, and his behavior is a reflection of this discomfort. Thus, a cat's behavior deviates from the norm—becomes bizarre—to communicate his anxiety.

If it's fleeting anxiety, which is quickly discharged, it's nonthreatening. The anxiety will disappear as soon as your cat is gratified. (Refer to the Emotions chart in Chapter 5.) But prolonged anxiety that builds up and is not immediately discharged can turn into pervasive fear or timidity that threatens a cat's peace of mind and physical well-being.

A cat is indeed a creature of habit. It's not unusual for a cat to be affected by a change of daily routine. The extent to which it affects his behavior is determined by his individual catsonality structure.

Self-esteem is the key element in feline catsonality. It's developed during kittenhood and reinforced in every phase of a cat's life. The essential ingredients are a fulfilling relationship with the mother cat, positive interactions with littermates, and a stable, loving, cat–person relationship. From these the cat acquires a sense of security and a good feeling about

himself. If any one of these ingredients is missing or negative, there's a chink in his self-esteem. He becomes vulnerable to ongoing anxiety and stress. It is the cat's reaction to stress that manifests itself in his behavior patterns and affects his catsonality.

A change of environment, the onset of sexual maturity, new companions, and separation from people or companions can cause a cat to have an uncomfortable reaction. If a cat's catsonality structure is immeasurably lacking in self-esteem, the reaction can trigger acute and long-term problems. Such problems may be manifested by aggressive behavior toward the cat's companions and/or persons. This is neurotic behavior and will continue or re-occur if the so-called blocks are not removed so that new coping mechanisms can be integrated into the cat's behavior. If the problem can be isolated and defined at the onset, therapeutic measures can release the aggression within a short time.

Preventive therapy is the best way to deter distress that can trigger aggression. Here are some recommendations:

1. If your young cat or kitten has started to bite indiscriminately, and he is your only cat or kitten, he may be a victim of the single-cat syndrome. A new cat or kitten will enable him to work out his energy in a healthy cat-like way. You might even consider a dog as a new companion. However, be sure to refer to my methods of introduction (Chapter 6) for a happy and successful bond.
2. If, when you entertain, your cat becomes anxious and glares at either you or your company, mention your cat's name in the conversation and take some time to stroke him gently so he doesn't feel rejected. Although a cat does not understand a person's words, he's very receptive and sensitive to body language and tone of voice.
3. If your cat is extremely antisocial, it might be best to provide him with his own private sanctuary when you have a gala party. However, escort him to his retreat cheerfully so he doesn't feel he's being punished.
4. If your cat displays aggressive behavior to you or his companion when you've been on holiday or away from home excessively, try to forestall such behavior by making arrangements for someone to look after your

cat if you will be gone overnight. Otherwise, leaving a voice message on your machine should be sufficient to give your cat a positive and wanted feeling should you be out for several or more hours.

5. Catnip is often a way for your cat to work off any pent-up energy. It generally produces a mellow response.
6. Try to spend at least ten minutes of daily quiet time with your cat, establishing a flow of relaxation between the two of you.

Corrective therapy is necessary when a cat has a long history of aggressive behavior. Such behavior is usually manifested by inter-cat hostility and/or attack cat tactics toward people. Such a catsonality is usually beyond neurotic and is perhaps psychotic.

The following recommendations will help you to modify your aggressive cat's behavior so he becomes a "reformed" attack cat:

1. If you have a single older cat—F.A.S.P.—Feline American Solitary Prince or Princess, use my method of introduction to bring in a new *kitten*. It would be harder to bring in an older cat. (See Chapter 6.)

   Sheba was a young spayed cat whose person, Wendell, was victimized by Sheba's aggressive behavior. She especially engaged in her vigilante attacks whenever he left his apartment. He had originally found her outside his bookshop and knew little of her past history. Sheba's cure was a young male kitten whom I personally introduced. Sheba now has constant company. No more anxiety attacks when Wendell departs! She's indeed a reformed F.A.S.P.!
2. For inter-cat hostility, refer to the aggressor's victim as his cat—such as "Tiger's Whitey"—to create a positive bond and defuse the jealousy complex.
3. If a particular person is your cat's tension target, speak of that person as your cat's person and your cat as that person's cat.
4. Audiotapes or CDs are effective in helping to relax your cat and prevent incidents.
5. Sometimes tranquilization therapy is necessary as auxiliary support in conjunction with a personalized support regimen of relaxation

techniques. It is unwise and ineffective to treat such behavior with just a tranquilizer. A tranquilizer treats only symptoms.

The support regimen is the primary treatment. It enables the cat to increase his stress tolerance on a long-term basis. As a cat's stress tolerance increases, and "incidents" decrease, he can slowly be weaned off the tranquilizer, but his support regimen should continue. Alternative methods such as homeopathic remedies, Chinese herbs, and acupuncture can also be used as auxiliary support.

6. The more relaxed you are, the less anxious your cat will be. Any exercise or activity that you can do at home to relieve your tension will nurture your cat. You can relieve your cat's anxiety and prevent an anxiety attack with the aid of my Emotions chart (Chapter 5) and by being good to yourself. Stress reduction for yourself can prevent catastrophic behavior in your cat.

chapter eleven

travel and care when you're away

"It beats me!" exclaimed Don. "I left my cat at the vet when I went away, and now she comes home sick and angry."

Like so many other people, Don boarded his cat at the vet while he was out of town, thinking that he was leaving her in a safe place. He sought my advice as to why his best-intended efforts for his cat had led to disaster, and what alternatives were possible for the future.

I explained to Don that a cat is prone to extra emotional stress when his person is away, and that this stress can lower his resistance to disease or trigger a medical problem. A cat's vulnerability to sickness only increases when he is kept near other cats. His body becomes tense, contracts, and is an excellent target for a stray virus or bacteria.

Don still couldn't understand why his cat was angry. I told him that she felt he had rejected her when he went away. In addition, she had probably felt threatened and insecure in an alien and cat-filled environment. She discharged her pent-up anxiety in anger.

I added that although many veterinarians will board cats, they do it as a service for their clients, in particular for cats that need daily medical attention. However, I don't think it is usually prudent to board your healthy cat at a hospital.

Don asked where it would be best to board his cat. I answered that, although there are exceptions, boarding is not a good choice for a cat for the reasons I just mentioned. I recommended he have a friend stay with his cat while he was away, or have a neighbor or cat-sitter come in a few times a day to care for the cat. If necessary, he could even leave his cat at a friend's house, preferably one without other animals.

Another alternative I do not recommend is simply leaving your cat at home alone for several days. This can have unforeseen negative results, as a client named Sara demonstrated when talking about her cat, Veronica.

Sara had always been a dog person. She had adopted a cat on impulse. Veronica, the cat she adopted, was over one year old, female, and spayed. All Sara's ideas about cats were secondhand. She had never before lived with a cat. Now she had a new surprise every day, and she had many questions.

"I'm going away for four days," said Sara. "It is so easy having a cat. All I have to do is leave lots of dry food, a large bowl of water, and fill up the litter box. Is there something else I should do?"

Sara's solution for her cat's comfort while she was on holiday was unfortunately a cat's nightmare. Sara had been influenced by another common cat myth: "A cat doesn't need or care about people."

Although cats who go outdoors usually don't need as much people contact, indoor cats are very much affected by interaction with their people. I explained to Sara that food and litter were only half the problem; although to start, four days of dry food was far from a well-balanced diet. Veronica would be affected by Sara's absence because she would experience stress from separation anxiety.

Sara immediately told me that once she had gone away for two days and asked her neighbor to look in on Veronica. When Sara returned, her neighbor remarked that she had seen neither whisker nor tail of Veronica when she came in to feed her. "You are sure you have a cat?" she teased. From this experience Sara concluded that Veronica didn't need anyone and preferred her privacy. So during her next weekend trip she left Veronica entirely alone.

I told Sara that a cat is a creature of habit and does not like a change of routine, because it presents conflict. Sara's absence was the first major change; her neighbor's appearance was another change. Veronica reacted by hiding, but even though Veronica chose to keep a low profile, she was aware of somebody coming in and checking on her. So she knew she wasn't entirely stranded.

Sara sighed and shook her head as she blurted out that maybe Veronica had had an adverse reaction to staying all alone during Sara's last holiday. A few days after Sara returned, Veronica started to lick furiously at her feet.

"Yes, that was definitely how Veronica coped with her feelings of anxiety," I replied. Very quickly Sara realized how much Veronica depended on her for attention and how Veronica felt when she was away.

The next time I spoke to Sara, she told me which friend of hers she had arranged to have Veronica stay with while she was away. Veronica knew and liked the friend, so it could prove an ideal solution.

If Your Cats Stay with a Friend

1. Pack their favorite toys, an abundant supply of food, their food dishes, litter, litter box, and scooper. Don't forget their scratching post!
2. Take along something that carries your scent—a bathrobe or T-shirt, for example. A few sniffs will make them feel secure.
3. Have your friend set up their litter box in the bathroom. Remove them from their carrier right beside the litter box, so they know where to take care of that business.
4. It is best to have them get acquainted with one room at a time, especially if the place is big.
5. Remind your friend to keep the windows shut or only slightly cracked so your cats don't take an accidental spill.
6. Leave written instructions about their diet, vet, and medical problems, if any, and leave the name of someone else who can look after your cats if your friend has to leave town.
7. Cuddle your cats and reassure them that you will be back for them.

Let your friend know that your cats may hide for a couple of days or that they may go off their food. But if they fast completely or stop urinating for

more than two days, the vet should check them out. If your cats are making their debut at your friend's place, it may take them a while to feel at home. But if your friend really makes them feel good, they will probably become bedmates.

If a Friend Stays with Your Cats

1. Be sure to have your friend visit a few times before you leave, so that your friend and the cats are familiar with each other.
2. Select a friend who not only likes your cats but has a mellow personality and a strong sense of responsibility.
3. Remind your friend to keep your cats on their usual feeding schedule. Within their own environment they will become confused and frustrated if their daily patterns are radically changed. You want to subject them to as little stress as possible.
4. Warn your friend that the cats may act a little strange for a couple of days; they may only nibble at their food. However, if their behavior remains bizarre for more than a couple of days, a visit to the vet may be necessary.
5. Remember to leave written instructions as to any special chronic medical problems, their diet, their vet, their favorite games, and how they prefer to be scratched and petted.
6. If your cats are on medication, it might be best to increase the maintenance dosage while you are away. Because of the additional emotional stress, they will probably need extra support.
7. Instruct your friend not to leave any windows open.
8. Leave your phone number and address in case of an emergency.
9. Let your friend know who can fill in if he or she has to relinquish your cats' care.
10. Give your cats a big hug and kiss, and tell them you'll be back soon.
11. If you are going away for only a couple of days, arrange for a friend, neighbor's child, or professional cat-sitter to come in at least twice a day to feed and visit with your cats. If you do engage a professional cat-sitter, be sure to check out the person's references.

Isn't it best to leave your cat in his own familiar environment if you have to travel?

It is, if you have someone who likes cats who can either stay at your place or visit and feed at least two times a day.

But my cats respond only to me, and they have a good relationship with each other. I have always left them with a lot of food and water when I have been away for a couple of days. I know they wouldn't even respond to someone else.

Perhaps they wouldn't respond, but they still need the presence of a person to look after their needs, such as fresh food and their litter box. Even if they choose to ignore the person, they are aware of positive attention. Also, if one of your cats took sick, the person could contact you, if possible, and take your cat to the vet.

But I don't have any friends or neighbors who could come in and look after my cat while I am away, and there isn't anyone with whom to leave her.

Then you should contact your vet to see if he knows of any professional cat-sitters; he may have some employees or clients who are available to help out. Or you may locate a senior citizen or a young person who particularly enjoys cats and make an arrangement with that person.

I have to go on a business trip for three weeks. Would it be best to have someone come in each day to look after my cat, or should I leave him with a friend?

Since you are going away for quite a while and you have only one cat, I would recommend that you leave him with a friend. If the time period were less than two weeks, it would be all right to have someone visit your cat at home.

My cats go outdoors, and when I have to travel, I always have a neighbor feed them and check up on them. Is there anything else you would recommend?

Tell your neighbor not to let them out after dark. It is best to keep this in mind at all times, but especially when you are out of town, because there is more chance that they will run into trouble when it is dark. By feeding them at sunset, you can get them into the habit of coming home early.

I have to leave town for a couple of weeks, and my friend, who has several cats, has offered to have my cat, Suzie, stay with her. Would it be a good idea to accept her offer, or should I have someone stay with my cat?

If your cat isn't used to other cats, it is not worth subjecting her to the stress of interacting with them, unless your friend has a large house where she can keep Suzie separate. Suzie would do best at home on her own turf. If someone can't stay with her, arrange for someone to visit each day.

I know my neighbor will look after my two cats when I go away, but I hate to ask him to change the litter box. Any suggestions?

Purchase some plastic liners that fit in the litter box to hold the litter. When the box has to be scooped or changed, your friend can simply remove the liner and litter together, and replace it with a new liner and fresh litter. You could even put down an extra litter box, alongside the other one. If you still feel your neighbor would be uncomfortable with these arrangements, you might engage a professional cat-sitter.

Traveling with Your Cat

What's your opinion about traveling with your cat?

I think it is a fine idea as long as you make the necessary provisions and as long as you will have time to spend with your cat when you reach your destination.

What about traveling with an older cat?

If your cat is up there in age, is an indoor cat, and has a high-strung personality and/or any chronic medical problem, it would be best for him to remain in his own environment.

My husband and I are planning an automobile trip to the country for a few weeks, and we want to take our cats with us. Is there anything we should know to make the trip more comfortable for all of us?

Your cats may need sedation if the automobile trip lasts four hours or longer. An oral or injectable medication may be used. (Injectable is usually more reliable.) Consult your vet for advice and for the sedative. Give your cats a trial dosage of the sedative beforehand, to make sure it will be effective, especially if you plan to use oral medication. Dramamine sometimes helps if your cats have nervous or jumpy stomachs. (Again, consult your vet for proper dosage.)

What about a rabies vaccination and other vaccinations?

A rabies shot may be needed if your cats will be going outdoors. Check with your vet to find out if it is essential for the particular area and if any other vaccinations are needed.

Should we feed the cats before we leave?

It is best not to feed them for at least four hours before departure. Their stomachs will do better empty than upset, and I'm sure you'll want to avoid a possible vomiting bout.

What about water?

They should have water available up until departure, and you might take a container of water and bowl along with you, in case they need a drink during your trip.

What supplies do I need to take my cats on an automobile trip?

You will need a disposable cardboard litter box, or a plastic pan, and plastic litter box liners. They can be purchased in most pet-supply departments of

supermarkets, dime stores, and department stores. You will also need the cats' litter and scooper, food, bowls, catnip, scratching post, and favorite toys.

One of my cats is on medication for his bladder. Should I increase the dosage before or during our trip?

Generally it is wise to increase the maintenance dosage during times of stress. Traveling usually is a stressful time, since it is a change of routine. Check with your vet.

What is the best way to carry our cats to the car?

Put them in their carrier, which should be lined with strips of newspaper in case they have an accident. I prefer the kind with the wire top, because it allows the air to circulate more freely. Make sure you have your identification on the outside of the carrier.

I am not sure my cats will enjoy traveling within the same carrier.

If they prefer privacy, take two carriers. You don't want them to get into a squabble before you even arrive at your destination.

Suppose they take cover when it is time to leave. What then?

Lure them out with some catnip or a morsel of food, and relax and breathe deeply so your tension doesn't send them into hiding.

How should I set up the automobile so the cats will be comfortable during our trip?

Arrange the litter box in a convenient area of the car. Try to keep the car cool and airy. Most cats prefer to stay inside their carrier, because they feel secure and protected. But if your cats have free range of the car, make sure the windows are open only a crack.

Your cats may become frightened by traffic noises and disoriented by the car's motion and wide-open spaces. Reassure and comfort them so that they relax and calm down.

Park in a shady spot if it is a warm day, and leave the windows open a crack if you have to leave your cats alone in the car for a while. Don't subject them to heat prostration! On a clear sunny day when the outside temperature is 80 degrees, the temperature in a closed car can easily rise to 120 degrees.

Do you think their behavior will be unusual when we reach our country spot?

Your cats may talk nonstop or hide upon reaching your destination. This behavior may be delayed for a few days if they have been tranquilized. Your cats will probably adapt to their new environment within a few days. The following instructions will make your cat's arrival less traumatic for all of you:

To Ensure Calm Arrivals

1. Arrange the litter box and water bowl in the bathroom.
2. Set up the carrier beside the litter box, and close the bathroom door.
3. After a few minutes, open the carrier so your cat can familiarize himself with the litter box and bathroom.
4. Leave the carrier open a while, so your cat has a security blanket containing his scent.
5. Stroke him, repeat his name softly, and hug him if he is anxious.
6. When he appears familiar with the bathroom, he is ready to conquer new frontiers.
7. Let him explore one room at a time if the place is large, so he doesn't become overwhelmed by the strange and vast space.
8. A pinch of catnip might be a welcome treat. You can feed him a light meal about four hours after arrival.

It will take a few days for your cat to adapt to his new environment, and he may even decide to hide out for a while until he feels braver. However, if he hides for more than a few days and is not eating or using his litter box, I would contact a nearby vet. You want to be sure emotional stress hasn't triggered a medical problem.

Do you think it will be okay for our cats to go outdoors at our vacation spot?

Yes, but there are precautions you should take beforehand.

Basic Hints for Meeting the Great Outdoors

1. Your cat should be completely familiar and comfortable with the inside of your place before you allow him to venture outdoors.
2. Even if he is tattooed or has a microchip, provide him with a collar and identification. A flea collar may be needed if he isn't already on an anti-flea pill.
3. It might be beneficial to install a commercial kitty door, if you own the place at which you are staying.
4. Don't allow your cat to go out alone at first. You should accompany him on short strolls outside the house. A kitty harness and leash would be useful. Your cat must first familiarize himself, under your supervision, with the immediate area surrounding your place. Then, when he is on his own, if he panics for any reason, he will know where to hide or how to return inside quickly.
5. Your cat should return indoors before dark. Lure him home in the evening by feeding him dinner at sundown.
6. You might want to look into the enclosure made by Kittywalk Systems, which allows your cat to enjoy the outdoors in a safe and controlled environment (refer to my website).

What about when it is time to return to the city?

Don't forget your cats' sedation, if it is necessary, and make a checklist of their supplies, so you don't forget anything.

Have you had any experience with taking your cats to hotels?

Yes, I have hotel-hopped many times with my cats. Many hotels permit cats in their rooms (although they may ask you to sign an agreement holding you responsible for any damage your cats may cause) and some are particularly cat-friendly, both here in New York and across the country. I have stayed at the Ritz Carlton in Montreal, a cat-friendly hotel, where I have even arranged and seen appointments. (Refer to my website for more information.)

Is it best for us to carry our cats to our room when we arrive at the hotel?

Yes, but if your cats' carrier isn't easily identified, let the bellhop know which bag your cats are in—if they are the silent type. If they start to cry, talk to them softly and calmly. The more excited you are, the more vocal they will be.

Any special instructions we should follow when we arrive at our room?

The same instructions apply here as were listed for arrival at a vacation spot. Also, be sure to put up the "Do Not Disturb" sign whenever you leave your room. You want to be sure no one enters your room and accidentally lets your cats out. In addition, inform your maid that the cats are with you, and instruct her not to leave your door ajar when she cleans.

What about when we are ready to depart?

Make sure the room is spic-and-span for the next cats, and follow the instructions for departure in the series of questions about automobile trips earlier in this chapter.

What is your feeling about traveling with cats to another state or to a foreign country?

It can be done successfully. Some states and most foreign countries require various vaccinations for cats before entry. Consult your vet for information listing domestic and foreign requirements, including any quarantines. You can contact the embassy of the appropriate country for information regarding your cat's entry.

Travel By Air

What do I need to know if I travel with my cats by air?

Airlines can provide you with regulations regarding foreign restrictions and necessary forms, but the consulate of the country you are traveling to would be your best source. Arrange to do your checking in advance.

Which countries have quarantines?

England used to have a six-month quarantine, but now there are alternatives. Check with the British Embassy or with your vet. There are some islands of the Caribbean, however, where cats are not permitted at all.

Do I need a special cat carrier?

Many airlines have regulation carriers, generally made of plastic, which they require you to purchase from them. Whichever carrier you use, be sure to line it with strips of newspaper. Label the carrier with complete identification. Don't forget to include your destination. The Sherpa Bag is the bag of choice of the airlines (check my website for information.)

Can cats travel in the cabin?

Some airlines allow at least one animal per cabin, but the carrier must fit safely under the seat in front of you. Some passengers buy a seat for their cat or dog.

Can two cats travel in the same carrier?

Yes, sometimes, if they can fit comfortably, but check with the airlines. You should try to make your reservations as soon as possible, so your cats won't have to travel as "excess baggage" in the cargo or luggage hold. Some airlines even have a limit to the number of animals in the cargo hold.

Isn't it unsafe for animals to travel in the luggage hold?

In the past there were many casualties among animals that traveled in the cargo holds of jet aircrafts. However, now most luggage holds must be air-pressurized in order to carry animals. Check with the airline to confirm that its hold is pressurized before planning to stow your cat. But you can't be 100 percent positive that everything is under control. Before your cat even boards the plane, he can be affected by temperature changes, engine noises, and exhaust fumes. If the cat becomes overheated or chilled, or overly stressed, the cargo hold can only contribute to his discomfort—no matter how well it is equipped.

What will my cat be affected by once he is settled in the hold?

Heat, humidity, and restricted ventilation. If the jet remains on the ground for a long time in hot weather, the hold will become warm. Your cat's breathable air will be drastically cut down if the hold is overloaded with baggage. Hyperventilation or heat prostration can prove fatal.

How can I check into these factors to make sure things will be satisfactory for my cat?

If you get the feeling that the airline's ticket agent is indifferent about your cat's welfare, check with the cargo manager. If the response is the same, book your flight with another airline.

Hints to Ease Air Travel for Cats

Here are some specific suggestions that should make your cat's air trip easier and more comfortable:

1. It is best to reserve a direct flight so there is no possibility that your cat will be transferred onto the wrong plane.
2. Avoid weekend or holiday flights when you can. Your cat's handling is apt to be more cordial if the airline isn't pressured.
3. If the weather is warm, try to reserve a flight that leaves in early morning or after sundown.
4. Call and ask for a ground-condition report before leaving for the airport. If the temperature is 80 degrees or more at the airport, and the humidity is high, try to reschedule your flight.

5. Arrive at the airport at least forty-five minutes before takeoff.
6. Don't allow the reservations clerk to put your cat's carrier on the baggage conveyor belt if he is traveling in the baggage hold. You should carry your carrier to the boarding area and have an airline employee personally see that your cat boards the plane. Make sure the employee understands that he should delay boarding your cat until no more than a half hour before takeoff.
7. When you arrive at your destination, ask an airline employee to bring your cat to you. You don't want to have your cat waiting on a loading dock or in a concrete waiting room in inclement weather.

Can cats travel by railroad?

Some smaller railroads allow your cats to accompany you into the car as long as they remain in their carrier. If a railroad has a baggage car, it will usually accept cats. You can find out this information when you make your reservation.

What about cost?

It usually depends on which class you are traveling. Any charge is usually determined by the weight of your cats and carrier.

What about weather?

Most railroad baggage cars aren't heated, so it wouldn't be wise to take your cats on a lengthy trip during cold weather in a northern climate. Also, avoid traveling by railroad with your cats in very hot weather.

How do I care for my cat during a railroad trip?

If your cat travels in the baggage car, often the railroad's personnel will water your cat at station stops along the way. Also, passengers are usually allowed to feed and visit their cats at station stops lasting ten minutes or more; at such times you can tidy and clean the inside of the carrier. If you travel on a train that allows you to carry your cat on board in a carrier, talk to him softly and repeat his name to make him feel secure.

Can cats travel on a bus?

Bus travel for cats is quite limited. Many major interstate bus lines in the United States don't permit cats. The same applies to some intracity lines. However, it may depend on the whim of a particular bus driver.

What about travel on a cruise boat?

Most cruise liners don't have provisions for animals. However, some have kennels. Check with the particular liner if you are contemplating a cruise with your cat.

Keep in mind that if your cat has a medical problem, the stress of traveling may complicate it. Discuss this with your vet. Generally a tranquilizer can do much to relieve your cat's emotional stress.

After you return home, don't be surprised if your cat is disoriented all over again. But don't fret! He will settle in quickly. Each holiday with your cat, especially if it is shortly after the preceding one, should get easier. One day you might decide to take him around the world!

chapter twelve

introduction of a new person

"I'm so upset," cried Betty. "My doctor told me that it wouldn't be healthy to keep my cat, Spotty, around after my baby is born!"

Betty was not the first of my clients to have this problem. Her physician's attitude about cats and newborn babies is one I often encounter. Many a medical doctor thinks that the mere presence of a household cat is unhealthy for a baby. However, this is rarely true. Betty was also influenced by another infamous cat myth—that a cat can steal the baby's breath away. But a cat cannot "steal a baby's breath away" unless the cat were to sleep lying across the baby's face, and the baby were too young to move or roll out of the way. I told Betty that a cat and a newborn baby could coexist healthily and happily in the same household.

But as Spotty was unspayed and already had experienced a few heats, I told her it would be best to have Spotty spayed before her baby's arrival. I also advised that Betty talk with her husband and work out a plan whereby he began to pay Spotty extra attention before the baby arrived, so the cat wouldn't experience a drastic change when Betty's time was primarily taken over by the baby. As long as she and her husband remained calm and comfortable about Spotty and their baby-to-be's relationship, all should go well. (See section in this chapter, "How to Introduce Your Baby to Your Cat.")

Why won't an unspayed cat do well with a baby?

An intact female is usually particularly vulnerable to anxiety reactions, and the baby's high energy level could precipitate such reactions.

Isn't there a disease a pregnant woman can catch from her cat even before the baby is born?

You are thinking of toxoplasmosis. Anyone can contract this disease, which is caused by a microscopic organism called Toxoplasma gondii. The infective organism is found in cat feces, but it is also common in uncooked beef, many cases occurring in people who eat steak tartare. The symptoms are generally a sore throat and possibly swollen lymph nodes in the neck. Many people have an immunity to this disease.

What is the danger to the unborn child?

The potential danger is that the mother will contract the disease during the first three months of pregnancy. The organism can produce cysts in the brain of the fetus, which can result in mental retardation.

So how can a woman prevent exposure if she is pregnant?

Her husband or child can clean the litter box.

How can a woman tell if she is susceptible to toxoplasmosis?

The best way is to have her physician take a blood sample and have it tested for Toxoplasma titer. If the titer is positive, she has already been exposed and is now immune. If the titer is negative, then she should avoid the specific sources of infection.

Why can a cat get upset by a baby?

The energy level of a baby is very high, and babies move abruptly and make lots of sudden noises. These can disconcert a cat who is used to a quiet, peaceful environment and even cause the cat to become threatened and anxious.

What will the cat do then?

To protect himself, he may withdraw from the baby or, if he is in close quarters with it, he may strike out.

If he runs off and hides, that is one thing, but how can you keep a cat from striking out at the infant?

If you know your cat is high-strung and aggressive, a tranquilizer would help to relieve his anxiety and calm his behavior.

But I don't want to tranquilize my cat.

Then I would advise you to keep your cat in a different room from your baby when it is crying or noisy.

My baby is due soon, and we have a very calm tomcat. Do you think he will accept the baby's arrival without problems?

It would be best to have your tom neutered before the baby's arrival. A tomcat's energy level can be very high at times, and the baby's arrival might transform your calm tom. In fact, it might cause him to spray the baby's belongings and become quite aggressive.

Why would this happen?

The baby's high energy level could easily bring out your cat's macho tomcat characteristics.

Suppose I let him go outdoors, so he can exert this energy sexually?

That will help if you don't mind treating him for battle scars from fights with other toms, but he may still be difficult to live with. Also, he is then adding to the ranks of unwanted kittens. (Refer to Chapters 14 and 15, Sex and Breeding.)

Can't a cat carry infectious diseases to a baby?

There is a far greater chance that people will transmit diseases to the baby.

What about ringworm?

Ringworm is a potential problem only if your cat shows signs of infection himself. If he does, then have your cat treated by a veterinarian, and wash your hands thoroughly after handling your cat. (See Chapter 8 for a detailed description of ringworm.)

How do you feel about a cat's sleeping in the baby's crib?

It is most likely that a cat will crawl into a baby's crib when it is empty. He becomes intrigued by the new smells, enjoys the security of a closed-in, comfortable spot, and wants to share the attention that the baby receives when in the crib. If you feel nervous about your cat's sleeping with your baby, put a screen over the crib so the cat can't enter.

I plan to breastfeed my baby. Do you think my cat will react to this?

Don't be surprised if he stretches out nearby and purrs. Your cat can pick up the happy, fulfilling feelings that nursing creates.

How can I keep my three-year-old from mishandling our cat?

When your small child mishandles your cat, you could try telling the child why the action is uncomfortable, and demonstrate by touching him in the same manner he touches the cat; e.g., show him how uncomfortable it is to be yanked up by an arm.

Dexter, Faith, Fheytan and Difi are a family of four cats who will soon be joined by a pair of adopted human twins from Turkey. Their guardians have provided the twins with a robotic cat in preparation for their arrival. The cat lets out a hiss whenever they pull his tail. This is what I call "creative prevention."

Do you know of many cases where a cat and baby couldn't happily coexist and the cat had to go?

I am reminded of a particular case told to me when I was interviewed by a writer for Glamour magazine. She had a very temperamental female Siamese cat who walked a thin line between acceptable and unacceptable cat behavior. The writer's new baby was not in her cat's script. The cat became

very irritable and very antisocial to the point where she behaved very aggressively to the baby. The writer didn't know how to deal with her cat's behavior. She knew her cat was acting out of self-defense but that the situation couldn't go on.

She found a new home for her cat with two older ladies who spent a lot of time at home and could provide the cat with all the care she would ever need. Although there can be cat vs. baby problems, the problems are readily solved if the cat's people are willing to take the time to work them out.

Do you think a Siamese cat is more prone to such behavior than a domestic or "boulevard" (as you call them) cat?

Many Siamese are very high-strung and people oriented; their behavior patterns are therefore usually more noticed because of their expressive and penetrating voice and distinctive body movements, but they are no more prone to aggressive behavior than other cats.

What age cat or kitten is best suited to live with a baby?

Unless a cat has had positive experiences with a baby already, it is usually best to start off with a cat under four years of age, because his patterns are less fixed. On the younger side, a four-month-old kitten is big enough to

scoot to safety if he wants to retreat from the baby; for this reason, I wouldn't recommend a younger kitten.

Have you known of any cats whose personality has actually been improved by a baby?

I am reminded of a cat named Spunky, who was standoffish with people but fascinated and intrigued when a baby paid a temporary visit. She spent most of her time close by the baby, so she could observe the baby's actions. Her person told me that the baby had a very relaxing effect on Spunky.

What do you make of this?

I feel the baby brought out Spunky's protective and maternal instincts; she was not at all threatened by the baby.

Can the arrival of a baby cause a cat to have medical problems?

Yes, sometimes a cat's stress tolerance cannot cope with a baby's high energy level, and the emotional stress may trigger bladder, skin, or respiratory problems. (Refer to Chapter 10, Catsonality Problems.)

What is the best way to introduce a seven-year-old child into a household that includes an older cat who has never lived with children?

First, you should double the attention you give your cat, before and after the child arrives. Instruct the child on how to interact with your cat, and keep your cat separate for a while if the child has playmates over. Have the child take over feeding your cat, and let your cat adapt to the presence of children gradually. Don't overwhelm him or force him to interact with the child.

Honey, my eight-year-old cat, and I have always lived alone, but I plan to be married soon. Do you think Honey will have a bad reaction when my future husband moves in? She gets along fine with him now.

To keep Honey from developing a jealousy complex, make it a point to overindulge her with attention. You want her to feel that your husband is another source of attention to her, not depriving her of your attention.

My two cats get along well with visitors, but they have never lived with other people. What do you think they will do when my new roommate moves in along with her two cats? I have a large four-room apartment, and my roommate will have her own bedroom.

I would expect that your roommate and her cats will be a definite threat to your cats' daily routine. Cats are territorial, and by nature will object to sharing their territory, especially with other animals. Your cats will be doubly threatened if you show any attention to the new cats. To avoid major spats, keep your cats separate from the newcomers for at least a few days, and then slowly bring them together. (See section How to Introduce a Roommate and Cats into a Household, pages 197–199.)

When a new person joins your household, you want to make sure your cats don't feel that this person stands between you and them. You want them to feel that the new person is a positive addition who will benefit them.

How to Introduce Your Baby to Your Cat

If your cat has never lived with a baby, you can make the encounter easier by following these pointers:

Before the Baby's Arrival

1. Try talking about the baby to your cat, using the cat's name so he doesn't feel left out. But be casual, not sugary or manipulative or your cat will become resentful. Don't say, "You'll love the baby." Instead, "There's so much you can teach the baby." Naturally, your cat won't understand your words, but will be affected by your body language and tone of voice.
2. You might sprinkle some baby powder or oil on your skin so your cat becomes familiar with the scent associated with the coming baby.
3. Prepare a high perch for your cat so he can hang out there whenever the energy level from the baby gets too disturbing.
4. Allow him to sniff the baby's furniture, etc., so his curiosity is satisfied.
5. Make sure his favorite foods and treats are on hand for when the baby arrives.

6. If your cat is sexually mature, have him or her neutered at least two weeks before your baby's arrival.
7. Arrange to have a neighbor's child come in and spend time with your cat if you find your time is full. The child can continue to come after the baby arrives, too.

One of my clients planned to adopt a baby and wanted to avoid stress for her senior cat, Ned. I advised her to record the sound of a crying baby and play the tape frequently so that Ned could familiarize himself with the new sound. My other recommendation was to stroke him and talk to him softly, while he listened, in order to reinforce a positive association.

The Baby's Arrival

1. Your cat should be fed when the baby arrives, so he has a positive association with the baby.
2. Permit him to smell some of the baby's belongings so he gets acquainted with the baby's smell.
3. Try to let him sniff the baby when the baby is relaxed, so he picks up a mellow feeling from the baby.
4. Try to make sure your cat gets his share of attention and, if possible, extra attention.
5. You might have a neighbor's child play with your cat if you feel the cat is neglected.
6. If you are worried that your cat will jump into the crib with your baby, you can put a screen or netting over the crib until the baby is old enough to move and roll over. Installing an intercom will also allow you to hear any sounds from your baby, or in the room.

There is every chance that your cat will adopt baby as his best friend or, if he is baby-shy, that he will keep at a safe distance.

How to Introduce a New Person into the Household

So your cat doesn't feel that a new person will deprive him of attention and contact from you, keep the following pointers in mind:

Before the Person Moves In

1. Have the person bring your cats treats and toys a few times before moving in.
2. If the person feeds him, your cat will form a positive association with the new person. Or try to feed your cat whenever the person visits.
3. Don't force the new person on your cat, or your cat may choose to be standoffish. Remember, a cat likes to make his own choices.
4. If your cat sleeps with you, and the person is going to share your bed, let the cat become familiar with the scent on an article of the person's clothing.

The Move-In Day

Don't schedule any house parties that day, and plan to spend most of your time at home, so your cat doesn't associate the new person's arrival with your departure. You want your cat to think of this new person as a source of additional attention and not as a source of competition.

Keep in Mind

If within a few weeks your cat doesn't accept the new person, it might be a good idea to adopt another cat as a companion for him. However, if you do, be sure to introduce the cats to each other carefully. (See Chapter 6, Choice and Introduction of a Companion for Your Cat.)

How to Introduce a Roommate and Cats into a Household

Before the Arrival

1. Have your future roommate over to visit a few times.
2. Don't allow the roommate to interact with your cats too much, as you don't want your cats to feel betrayed when the roommate moves in along with other cats.

3. Give your cats some toys that smell of the other cats, so they will become familiar with the new cat smells.
4. Decide on which area (ideally, the roommate's bedroom) the new cats can stay in, apart from your cats, for the first few days after their arrival.
5. All the cats should be neutered if they are sexually mature—at least two weeks before the introduction.

Day of Arrival

1. Plan to feed your cats shortly before or at the time of arrival, so they form a pleasant association with the newcomers.
2. Have your roommate place his or her cats in their designated area, and be sure to shut the door.
3. After the newcomers are out of their carrier, take the carrier to your cats, so they can explore the inside and get a good whiff of the newcomers' scent.
4. Return the carrier to the newcomers, so they can become familiar with your cats' scents.
5. During the next few days your cats will probably hang around the newcomers' door. This is okay,, because it will enable the cats to acclimate to each other's smell.
6. Do not enter the newcomers' room, or your cats will become suspicious and jealous.
7. Dote on your cats, and your roommate can dote on hers. Don't try to play round robin with your attention.
8. After a few days have passed, put your cats in the newcomers' room and the newcomers in your cats' territory, so they can all check out the terrain. But make sure the door is closed between them and that they remain separate.
9. Repeat this periodically for a few days.

Introduction Day

1. By the end of a week the meeting between all the cats can take place. Pick a non-hectic day, one on which you and your roommate will be comfortable and relaxed.

2. Before you open the newcomers' door, have food prepared for both sets of cats and place it where they will find it when they enter each other's territory. They may not be interested in it, but they should have the option.
3. Don't monitor the cats' actions, and do all you can to stay calm and collected.
4. If there is a tussle, use the plant sprayer to cool off the cats.
5. Remember not to make any overtures toward the newcomers, and have your roommate avoid your cats.
6. Scoot them back to their respective areas if there is unbearable bickering, but there will have to be some aggressive interplay to establish roles. Your cats may be more aggressive, as it was their domain first.
7. If all goes fairly smoothly, don't continue to keep them separated. If not, repeat this procedure for the next couple of days until they work out a living arrangement.

Conclusion

All should go well if you devote your attention to your cats. Don't feel sorry for the newcomers and try to make them feel better by petting them, etc. Your responsibility is to your own cats. After all, the newcomers are a threat to your cats' territory and comfort, and your cats won't relax until they feel less vulnerable. The more secure they feel, the more receptive they will be to the newcomers.

chapter thirteen

all about grooming

Jim was all upset! He had just found a long-haired cat on his way home.

"He's such a mess! It looks like he hasn't washed or groomed himself in days," cried Jim. Up to now Jim had experienced only short-haired cats who kept themselves immaculate; he couldn't understand what was wrong with this cat.

I explained to Jim that when a cat is sick or upset, his coat reflects the problem. A cat frequently loses the desire to wash when his energy level is depressed. Most street cats, both long- and short-haired, don't have a dashing appearance, but a long-haired cat's fur suffers more from neglect. If a cat is on the run, keeping up with street dirt can be an unending battle.

Evidently Jim's new find had been out on the street awhile, and grooming was not a top priority with him—survival was. I told Jim that as soon as his new cat started to feel secure, he would begin to get his coat in order. I suggested Jim help by combing him and cutting out his mats. He could purchase metal combs and brushes of varying density and attack one section of fur at a time. He might even have to give his new cat a bath.

In general, to groom a messy long-haired cat, first you gently cut away his large matted patches of hair. Use barber's scissors, preferably, and always keep one blade next to the skin and cut carefully up through the hair to slowly detach the mat. Then brush the cat carefully and gently, moving down his body. Cats especially like being brushed around their heads and under their chins. Brush there to please them, and attack the coat only after you have got the big mats out. The key to success is gentle brushing that doesn't pull on the skin; and don't be too insistent about it.

Like many people, Jim didn't feel he could undertake such strenuous grooming efforts, so I recommended he call a professional groomer who would come to his home and do the job.

Normally a cat is fastidious about grooming himself and, often, his companions, too. Therefore it wasn't unusual for Jim to be startled and confused about his new cat's appearance. There are some cats who are lax when it comes to grooming, but generally if a cat is not grooming himself, he has a reason.

How often should you groom a cat?

You should try to make it part of your daily routine, especially if your cat is long-haired.

Do you use combs to groom a short-haired cat?

A rubber brush usually works better. Wet the brush first to keep the hair from flying around. However, if the fur is badly tangled, a comb may be the answer.

What do brushing and combing do for the cat?

Grooming removes the loose hairs and stimulates circulation, thus preventing the skin from drying up or flaking. However, if you notice an oily discharge near the tail, your cat's anal glands may need to be emptied by the vet. If your cat is a tom, this is often called a stud tail or discharge.

Suppose a cat doesn't like to be groomed. Is there any way to influence him?

Try giving him some catnip or a taste treat to relax him and please his stomach. The best time to groom him may be when he is sleepy and less likely to protest. Give him a treat when you are finished, so that he associates good things with being groomed.

My cat accepts a small amount of brushing, but then he objects. What can I do?

If his grooming tolerance is low, do a little at a time. Don't try to finish the job in one sitting.

I brush my cat once a day, but his coat is still dull and his skin is dry. Is there something else I can do?

Offer him an egg yolk a couple of times a week. You might also try butter. The protein in the egg yolk sometimes stimulates healthier hair, and the fat in butter may supply another deficient nutrient. Don't feed your cat raw egg whites, because they contain an enzyme that can destroy one of the B vitamins. However, it is okay to feed your cat cooked egg whites if he enjoys them.

Is there anything else that contributes to poor hygiene?

If a cat is fat, it becomes difficult for him to reach various parts of his body. The neglect can cause the skin to become dry and the hair to become matted.

Any other factors?

Poor nutrition, internal parasites (such as worms), external parasites (fleas, ticks), and any acute or chronic disorder will take nutrients away from the skin to support other body functions that are ailing.

What is it about a cat's tongue that enables him to do such a wonderful job of grooming?

A cat's tongue has a rough surface that is perfect for washing his fur. The action massages his skin and stimulates circulation; this brings more oxygen to his skin and enhances his coat.

My cat spends at least a third of his day grooming himself and his companion. There must be an added attraction!

The washing motion of the tongue makes a cat feel warm and relaxed. Comfortable feelings return from when his mother licked and preened him.

Is that why they say that when a cat is in doubt he washes?

Yes, generally when a cat is in doubt or conflict, he is anxious. The licking motion helps him to feel calm.

I think I know what you mean. When my cat, Ricky, is upset or wants to get my attention, he washes and licks his back leg until it is raw. How can I get him to stop this?

Sounds like Ricky is overdoing a good thing. The best way to get him to stop is to distract him by paying attention to him. If that doesn't help, and his licking persists, consult your veterinarian.

Why do you recommend having a groomer come to the home?

Because a cat is more comfortable in his own environment. He probably will become excited when he is groomed, especially if he is bathed. He is subjected to less stress if he doesn't have to travel to a strange place, where he is also a vulnerable target for germs from other cats. Anxiety can lower a cat's resistance. If the groomer comes to him, he will undergo less anxiety, and the health risk is smaller.

Suppose a cat's fur is too tangled for a groomer. What then?

It may be necessary for your vet to tranquilize or anesthetize him to get the job done. In very severe cases a cat may have to be shaved.

Won't that bother him?

Yes, he will be very embarrassed afterward and will need a lot of support and love from his person to ease his discomfort and boost his pride.

Are there any other major reasons why grooming a cat is so important?

A poorly groomed cat is a perfect target for emotional and physical problems, because when a cat's fur is all tangled and matted, he feels uncomfortable. He feels pain whenever he moves, and he becomes prone to skin infection. Things can get worse—his pain and depression can often precipitate urinary and respiratory problems.

But isn't it just the opposite—a physical problem can cause a cat to neglect his grooming?

It can happen either way. Unless you have a cat who goes outdoors and constantly rolls in the dirt, if your cat is neglecting his fur, chances are there is a problem. It could be that he has an emotional problem, that for some reason his self-esteem is low and he is short on pride. (See Chapter 10, Catsonality Problems.)

If you remember that a cat takes great pride and dignity in his appearance, you realize that any time he fails to do so, there must be a reason. It is important that you find out what the reason is, so the problem can be resolved as quickly as possible.

How to Give Your Cat a Bath

The more relaxed you are, the easier a bath will be for both of you. Your cat needs your whole-hearted support in order to remain calm. Ask a mellow friend to assist you if you need support yourself. Remember, breathe regularly and stay relaxed. If you are confident and calm, your cat is likely to be so also. If your cat is the nervous type and responds well to a tranquilizer, you may want to sedate him before a bath.

Materials Needed

1. A large kitchen sink or the bathtub
2. Adequate lighting
3. A mild shampoo, such as baby shampoo, and a gentle cream rinse or conditioner
4. Cotton swabs
5. Hair dryer
6. A few large absorbent towels
7. Metal comb or brush
8. Scissors
9. Garbage bag
10. Mercuric oxide ointment
11. Basin or pitcher full of water to use for rinsing
12. Plastic or newspaper to spread on floor around sink or tub
13. Ribbon (optional)

Procedure

1. Swab out your cat's ears (just the part you can see) with cotton swabs moistened with lukewarm water.
2. Apply mercuric oxide ointment around the eye area to protect the cat's eyes from soap.
3. Stroke and praise your cat in a gentle voice throughout the bath.
4. If your cat has long hair, start off by combing your cat's stomach area and underparts to locate and detach mats. Don't try to attack a huge hair tangle all at once. Separate it into smaller ones. Continue to talk softly to your cat as you comb, and breathe freely. Work from the outer tip of the

tangle in toward the skin. After you have separated all the tangles, and the combs go through the hair without a snag, your cat is ready for the bath.

5. Have the surrounding air at a comfortable temperature—not too cold—so that the wet cat won't get chilled.
6. Have on hand a pot or basin of tepid rinse water.
7. Place your cat in the sink or tub and hold him firmly but gently by the scruff of the neck. Wearing rubber gloves can protect your hands if the cat is the type that protests a bit.
8. Fill the basin with water. Do it slowly so there isn't a deluge of water. Wet the cat's back, legs, and belly thoroughly.
9. Apply the shampoo sparingly, so you don't have to rinse forever.
10. To rinse, use the pitcher or basin filled with tepid water, mixed with water from the tap. Next, apply rinse or conditioner. Once again, rinse thoroughly; if you don't, the cat may have a reaction to the soap on his skin and lick at his fur continuously.
11. Wrap your cat in a towel and rub briskly. Kiss and hug him as you rub.
12. Now he is ready for the hair dryer. Fluff his hair with a comb as you dry. Don't stop until he is dry and gorgeous.

13. If your cat won't tolerate a hair dryer, then towel-dry him as much as possible and place him in a warm spot (like a sunny windowsill or under a lamp) to dry.
14. If the cat gets chilled, warm him by wrapping him in a warm, dry towel or blanket, and hug him, using your body heat to warm him.
15. Clean the combs and give him one final comb.
16. Top off your work by tying a ribbon bow around his neck. He can take his frustration out on the bow if he is annoyed.

chapter fourteen

sex and breeding: hers

I knew it would be hard for Mrs. Davis to accept what I had to tell her about Devin, her two-year-old Abyssinian who had just aborted her litter. She had managed to give birth to one kitten the first time she had become pregnant, but the kitten had lived for only three days.

"What's wrong? Why can't Devin sustain a pregnancy?" Mrs. Davis asked me. "She is a registered Abyssinian with a magnificent lineage, and the Aby I mated her with is a splendid tom. I know her kittens would turn out to be beauties!" she exclaimed. "Isn't there something—anything—I can do to help Devin have a healthy litter?"

Looking at Devin, I could see that she was not only delicate in build but a shy and sensitive cat. Mrs. Davis confirmed my impression when she mentioned how Devin usually hid when people came over. She was surprised that my presence didn't cause Devin to withdraw.

When I arrived at Mrs. Davis's house, she was sitting on the sofa with Devin in her lap. She told me that her own children were now married and that her husband traveled a lot. Devin was her only cat, and she wanted so much for Devin to have a litter of kittens. She knew Devin would be a wonderful mother, and she herself wanted to be able to share in the enjoyment of Devin's kittens.

I took Devin's full case history, and then I offered my recommendations. I told Mrs. Davis that I fully agreed with her that Devin was a wonderful, beautiful cat, but there were other major factors to be looked at when considering breeding her again. Devin's pregnancy track record was poor; the strain and stress on her delicate body and exquisite sensitivity from these traumatic ordeals would certainly have affected Devin's health. Why take any more risks?

"But giving birth to kittens is such a natural experience for a cat," objected Mrs. Davis.

I told Mrs. Davis that although giving birth to kittens was indeed natural and beautiful for many cats, it wasn't right for Devin. Her previous experiences indicated that she lacked some of the essential ingredients for producing healthy kittens. Rather than try to figure out exactly what it was that Devin was missing or needed, I felt it would be best to realize that Devin was not meant to give birth.

"Well, what do you suggest that I do?" asked Mrs. Davis very sadly.

I answered that it would be best to have Devin spayed after she had recovered from her most recent incident.

"But won't spaying change her personality?" cried Mrs. Davis.

I explained that Devin had reached her full sexual maturity, and that therefore the surgery would in no way hurt her personality. If anything, she would become more relaxed and trusting. Mrs. Davis then volunteered that Devin's heats had always been severe, and that Devin had cried out a lot and seemed very uncomfortable during them. I reassured Mrs. Davis that once Devin was spayed, there would be no more painful heats for Devin to endure. Devin's experience demonstrated that she was not destined to be a mother and that further attempts would only seriously endanger her health.

Often the decision about breeding cats must be determined by the individual cat's health and temperament. Not reproducing is often much wiser, and a cat's life is in no way deprived just because it doesn't experience motherhood.

Isn't it necessary for a female cat to mother at least one litter before she is spayed?

No, that is another cat myth. A female cat should be healthy and sexually mature when she is spayed. It makes no difference whether or not she has mothered a litter. And with the present staggering population of unwanted kittens, there is no need to worry about a kitten shortage.

How is it possible to determine when a cat is sexually mature?

A male cat reaches sexual maturity between seven months and a year, but sometimes as early as five to six months. His coming of age is announced with the arrival of strong-smelling urine, which he may spray in unusual places, frequent crying near doors and windows, and rough, aggressive behavior. It is best to have him neutered before these tomcat characteristics become permanent habits.

A female can reach sexual maturity as early as five and a half months or as late as one year. She is ready to be spayed after her first heat. A female's first heat may be silent, but generally it's an extravaganza. She may become more vocal. Some females become extremely expressive and sound like frustrated lyric sopranos. She may become very sensitive along her lower back; don't be surprised if her rear shoots up into the air when you pet her there. You might expect her to crouch low to the ground with a low moan and to roll back and forth on the floor.

Why would she do this?

It helps to relieve the uncomfortable sensation she feels inside.

What other symptoms might I expect?

Sometimes a female urinates and leaves stool piles around the house to bring attention to her condition.

Sounds revolting! Can I expect it to stop?

Yes. Her normal toilet habits will commence several days after she goes out of heat. However, if she isn't spayed soon, her sporadic behavior may become a habit.

You mentioned that a female cat can reach sexual maturity as early as five and a half months and should be spayed after her first heat. Suppose she doesn't have an obvious heat. How can you tell when she has reached sexual maturity?

If your cat's heats are silent, it is generally safe to assume she has reached sexual maturity by eight months old. Some animal shelters now spay and neuter young kittens to prevent them from being bred or kept intact after being adopted.

Suppose she goes outdoors and before she is even eight months old she comes in contact with tomcats?

Then it would be best to make an appointment with the vet to see if he thinks it is time to have her spayed.

What is the exact name for the female spaying operation, and what does it consist of?

The operation is called an ovariohysterectomy. The uterus and ovaries are both removed, for if just the uterus were removed, the female couldn't become pregnant, but she could still go into heat.

Normally, the cat will have to stay in the hospital for one to three days. Her incision will be either on her side or, most commonly, on her abdomen. If the incision becomes red and swollen, or if she licks it excessively, you should have her stitches checked. Otherwise, once she is home, I advise giving her lots of attention, feeding her her favorite foods, and keeping her indoors if she is an outdoor cat. Within ten days the vet will remove her sutures. (See section "Your Cat's Postoperative Care" in Chapter 8.)

Will my cat go into heat after she is spayed?

No. However, if she does show symptoms of heat, she possibly has a retained ovary. If so, contact your vet, so he can decide when and if to perform an exploratory operation to locate the retained tissue.

Does this occur frequently?

No, but accidents can happen.

How will a female's companions react to her homecoming after the surgery?

There may be a little sniffing, but usually they soon settle down and everything is tiptop. However, the cat's feline companions may be threatened by her strange hospital scents and reject her because she smells like many different cats. There is a way to prevent this from causing problems. (See Chapter 2.)

I have heard a female cat becomes fat and lazy after she is spayed. Is this true?

After a female is spayed, she is more relaxed and her energy level isn't as high. Therefore, she may not burn up her food at quickly as before. If she doesn't go outdoors, her exercise may be limited.

So then she will become fat and lazy?

No, not if you decrease her food when you see that she is becoming too hefty. She won't need as much food as before she was spayed. Also, in many instances, a cat is spayed near the time she stops growing. If she continues to be fed as if she is a growing cat, she will become fat. There is little correlation with the spaying.

But everyone says a cat always knows when to stop eating.

That isn't fact; it's fancy! The large number of chubby to obese cats is proof.

And her exercise?

If she can't get enough exercise indoors or out, take her for walks in the hall. Chase her around and give her catnip to stimulate her to work out her energy. (See Chapter 3, Exercise and the Great Outdoors, and Chapter 7, Diet.)

So once a male or female cat reaches sexual maturity, he or she can be neutered without any effect on sensuality. Is that correct?

Once cats reach maturity, their sensuality is not dependent upon their sexual organs. They have already achieved the ultimate state of differentiation. At

sexual maturity the hormone levels trigger an area in the brain that, from then on, controls a cat's femininity or masculinity, without need of further hormone stimulation. Many emotional and physical problems are triggered by intact cats' sexual stress. Once a cat is neutered, a large source of potential stress is eliminated.

What is your general feeling about breeding a female cat?

If you decide to breed your cat, many important factors should be considered, such as your cat's basic body structure, general health, and resistance to disease. Also, you must ask yourself if you are capable of dealing with a pregnant cat, soon to become a nursing cat with kittens and all their disruptive energy and, ultimately, a mini–adoption center for kittens. Last but not least, can you assume all the economic obligations?

Suppose all these factors are in order.

Even if they appear to be so, there are still very painful, staggering statistics that shouldn't be overlooked: the millions of homeless cats that are put to sleep and the stray cats that often die a slow, painful death on the streets. Why add to this already huge population?

What if I line up homes for the kittens before they are even born?

Unfortunately, people will often agree to adopt a kitten one day and decide against it the next.

What if the female's an exotic breed and her type of kittens would be desirable?

If your cat is a particularly excellent specimen of her breed, you may want to breed her.

What is your feeling about aborting a pregnant female?

If she is under six weeks pregnant and is otherwise healthy, and you have a competent vet, I am in favor of aborting and spaying her.

Quite recently my friend found a pregnant cat outside her house. Her vet recommended that the cat be spayed and aborted at the same time. The gestation period is usually sixty-three days, and the vet estimated that she was only a few weeks pregnant. The cat's surgery went very well, but two weeks later, when it came time to remove the sutures, her breasts were terribly swollen. The vet predicted that in two weeks the swelling would disappear and her breasts would return to normal. He was right, but could you tell me why it happened?

Normally, removal of the ovaries and uterus eliminates the hormonal stimulation for breast development and milk production. But some cats have a paradoxical reaction and the process continues. As the vet said, the process usually lasts a couple of weeks and then the breasts return to normal.

Why do cats often get pregnant again immediately after having a litter?

Because they quite commonly go into heat while nursing, and if they come in contact with an intact male—voilà!—another litter.

Can a female be spayed while nursing a litter?

It is best to wait until two weeks after she stops nursing.

Taking Care of the Queen and Kittens

What recommendations would you make to prepare for a female cat who is about to give birth?

Line a sturdy box or basket with clean towels and set it up in a quiet, warm place. A drawer or the bottom of a closet is fine, if the cat indicates she has a preference for such a spot.

If I am home while she gives birth, should I stay by her?

Usually the female will let you know. She may cry out to you or attract your attention by running after you. If she wants your presence, remain calm and talk to her softly so she feels your support.

Suppose she has trouble while giving birth?

I would suggest calling the vet beforehand for some pointers and to find out whom you should call in an emergency if he is not at his office when she gives birth.

Does the vet often have to assist a mother cat when she gives birth?

No, hardly ever.

How long should the kittens stay with the mother?

It is best to keep them together until they are at least six to eight weeks old, unless her milk has dried up and she can't nurse them.

What is a good diet for a pregnant mother?

Increase her normal food intake by half during the last three weeks of pregnancy. Add a good multivitamin and mineral supplement. Calcium is very important for pregnant and nursing cats. Give her milk products or calcium pills (calcium lactate or dicalcium phosphate are two common supplements). After the births keep her diet the same or increase it, depending on the number of kittens and her weight. Some cats have been known to eat two or three times their normal dietary requirements while nursing.

If the mother's milk dries up, what can the kittens be fed?

A gruel of water mixed with canned or baby food and/or a commercial kitten milk supplement powder. And make sure you double the queen's food, multivitamins, and mineral supplements. It is up to the cat's person to separate the mother from the kittens before she becomes depleted.

What is the best time to start weaning the kittens?

At four weeks old they can start to eat solid food along with their nursing. If the kittens insist on nursing and refuse to eat regular food, there is no reason to rush the separation, as long as the mother is in good health. Otherwise, separate the mother from the kittens until they are ready to be adopted.

Can they all eat out of the same bowl?

Separate bowls are often best, so that the more timid kittens have a better chance of getting their share. The dominant kittens may try to push their timid littermates out of the way. You must monitor this. Be sure to keep a damp washcloth handy in case one of the kittens bathes in his food.

My cat's three kittens are three weeks old, and I have decided to keep one and find homes for the other two. Which one should I keep?

During the next few weeks, observe the mother and her kittens and see which kitten is most dependent on her. That would be the kitten to keep.

Should the other two kittens go to the same home?

Sure, if they are happy together, but if that isn't possible, try to find separate homes for them where another cat is already in residence.

I noticed that sometimes the mother cat retreats to an unattainable (for kittens) spot. Why?

She does this when she needs some time away from her kids. Don't worry, she knows when she can be spared.

I have them camped out in a small area in my bedroom where the kittens can't get lost. Does that sound okay?

Sounds fine. If the mother didn't approve, she would simply move her kittens to a spot of her liking.

Do you think my female will get upset when I give her kittens away?

I would be surprised if she were blasé.

What can I do to make her feel better?

If you keep one kitten, and the other two go to individual homes, try to arrange for them to go a few days apart, so your female's separation anxiety isn't as great. (See Chapter 5, Emotions.) Whatever you do, tell her days ahead and at the time of their departure that her kittens will be fine. She won't understand your words, but she will pick up your positive feelings

through the tone of your voice and relaxed body. Your calm feelings will help to relieve her anxiety.

Will my male cat hurt my female's kittens?

If the male isn't altered, I would strongly recommend that you have him altered very soon, even if he is the kittens' father. A tomcat's high energy level can often clash with the kittens' high energy and cause him to strike out at a kitten. There are exceptions, of course, but an altered male's stress tolerance is greater and he can adapt better to the new additions.

What can I expect from an altered male regarding kittens?

He may prefer to ignore the kittens, or he may be fascinated by them. Wylie, a client's cat who had always been very shy, blossomed when his people took in a mother cat with kittens. He became so enraptured by the kittens that he even tolerated their unfruitful nursing when the mother was unavailable.

My female had three kittens, and they were all so different from her in personality and disposition. Why?

A mother cat contributes only partially to her kittens' personalities. Her personality isn't always the dominant or sole factor affecting her offspring's

temperament; the kittens also acquire genetic characteristics from the father. There is often more than one father for any one litter, as the female can ovulate a number of times during heat and, if inseminated by a different male each time, can produce a litter with multiple sires.

My mother cat's milk dried up after four weeks. I had to give the kittens food supplements, and she paid little attention to them. Why did she act this way?

Evidently she didn't have the energy or the inclination to care for them. Luckily she wasn't out on the street and she had you to pinch-hit for her. Sounds like "mama-cat's lib" to me.

Will kittens use a litter box right away?

The mother takes care of their waste in the early stages. Later they will instinctively use a litter box. (See Chapter 9, Litter Box Problems.)

The Problems of Unspayed Cats

How often does a female cat go into heat?

Each individual cat has her own cycle. The most common cycle is twice a year.

Can a female's health be affected if she isn't spayed?

Yes, the unspayed syndrome can leave a female vulnerable to serious physical and emotional disorders. Recurrent heats are stressful and bothersome for both the cat and her person. Unless a female cat is bred each time she comes into heat, she won't ovulate (cats ovulate only after coitus), her hormones may become unbalanced, and she will become susceptible to illness.

What are the potential physical problems?

The two most common problems are cystic ovaries and pyometra. Ovarian cysts occur when a cat comes into heat and the follicle or egg produced in the ovary is not released. A pyometra is an infection of the uterus. These two problems can occur concurrently or independently.

What's the treatment?

Remove the ovaries and uterus!

Suppose there are males around?

If there are, she may eventually tire of the rough and tumble or may experience any of the many accidents that seem to befall female cats in heat.

Such as?

At The Cat Practice, I often encountered females in heat who had either darted out of the house and under a car or taken a skydive from a window. Tie-Dye was such a cat. She spied a tomcat from an open window and literally threw herself at him. It was quite a fall, and her passion brought her a broken leg. After that her person realized that Tie-Dye would be safer, calmer, and happier after she was spayed.

Also there were many, many abandoned females who were brought in either in heat or pregnant. Fortunately, we had a large number of sympathetic clients, and they would sponsor and find homes for the wayward females—if they couldn't adopt them themselves.

What emotional problems can be triggered by not spaying a cat after her first or second heat?

The emotional stress from her pent-up sexual energy can affect a cat's personality, sometimes dramatically. Because she is often uncomfortable and frustrated, her stress tolerance is lowered and she is easily threatened.

What do you mean?

If she becomes anxious or excited, she may lash out at her people or companions in self-defense. That is, if their energy level is too high and she can't cope with it, she may try to stop their activity by a sudden swipe, nip, hiss, or all three.

I still don't understand what you mean.

Let me tell you about Camelot. She was an unspayed three-year-old who lived with her person, Margaret. Camelot had always been a very high-strung cat. But although she was sometimes temperamental with other people, she was always affectionate with Margaret.

One day Margaret had a few people over for lunch, which lasted for a few hours. After they had left, Margaret was busy cleaning up when she noticed Camelot was nowhere in sight. She found her in the bedroom closet. But when Margaret went to reach for her, Camelot lashed out at her with her claws and hissed. She wanted nothing to do with Margaret. A few hours later Margaret again tried to make contact with Camelot. This time Camelot was more emphatic; when Margaret tried to pick her up, she sank her teeth into Margaret's arm. Margaret was aghast at Camelot's sudden personality change.

When Margaret called me for advice, she was certain that I would tell her Camelot was hopeless. She was relieved when I explained that Camelot was a victim of the unspayed syndrome. Camelot's high sexual energy was challenged by excitement and change. The festivity and interactions from Margaret's luncheon guests had threatened her and made her uneasy, and she withdrew so she wouldn't have to cope with the anxiety it caused her. She didn't intentionally attack Margaret; her actions were out of self-defense. Camelot couldn't deal with Margaret because at that moment she couldn't deal with herself.

I told Margaret that after Camelot was spayed, she wouldn't feel physical anxiety and frustration any longer, so her extreme tenseness and protective stance would vanish and she would be more trusting and easier to live with.

Margaret realized that Camelot would indeed benefit by the spaying surgery, but she was worried about one thing. She truly admired Camelot's sensuality, and she didn't want to take the chance of sacrificing this exquisite part of Camelot's being.

I reassured Margaret that Camelot's sensuality was not at stake. Since Camelot was sexually mature, her hormones had already triggered the area of her brain that controlled her femininity. Only if she were sexually immature when the surgery was performed would her sensuality be adversely affected. Camelot's behavior, like that of all cats, would continue to be controlled by her highly developed senses. Many cats become more sensual and affectionate after they are spayed, because they feel emotionally and physically more comfortable.

So what you are saying is that a female cat is apt to become Ms. Personality Plus after she is spayed. I can't believe there aren't any minuses!

The only minus I have encountered is when a female's person feels guilty about having had her spayed and treats her as though she is somehow deficient, such as when the cat's person refers to her as an "it" and belittles her.

But how can the cat understand what her person is saying?

The cat can pick up her person's tension by the sound of her voice and breath and by how she moves her body. This can affect a cat's behavior and personality. (See Chapter 10, Catsonality Problems.)

My friend has a female cat who is six years old, unspayed, and apparently very normal. The only problem is that she sometimes scratches and bites at her tummy. Although she never shows any obvious signs of going into heat, do you think this behavior is related?

I strongly suspect it is. She probably reacts in this way to the discomfort and anxiety triggered by her heats. It is her way of externalizing her condition. If she were spayed, this behavior would stop.

Could something else be bothering her?

There is a possibility that she may have an underlying respiratory ailment, such as asthma or a cardiac problem.

Why, then, would she bite and scratch her skin?

Her skin may be her secondary stress target, whereas her chest could actually be her primary stress target.

Well, do you think my friend's cat has a heart problem?

Not necessarily. Her heart may be perfectly fine. However, when she has her next checkup, a thorough physical examination and possibly chest X-rays might be in order.

I get the impression that a cat who has silent heats is sometimes a candidate for serious problems. Is that so?

Unfortunately, yes. When a female cat doesn't externalize her heats, it is very common for her person to see no apparent need to have her spayed. However, an unspayed cat is an ideal target for cancer because of the emotional and physical stress.

How does the cancer manifest itself?

Generally as breast tumors, discovered when the cat's person feels lumps in her cat's breast, which the vet subsequently diagnoses as benign or malignant.

What is the treatment?

Surgery and medical treatment, or an attempt to control the situation with medication. Either treatment should be combined with positive support therapy from the cat's person.

What's the prognosis?

It depends on the severity of the tumors. Some cats survive comfortably for months, others do well for years, but there are the unfortunate ones whose time is very short.

It seems like a person takes a gamble with a female's health by not having her spayed.

Absolutely! Unless a female cat is bred every time she has a heat, she can endure an immeasurable amount of emotional and physical stress. This can all be avoided by having her spayed when she is sexually mature.

I found a female cat several months ago, and the vet thinks she is about two years old. He checked for an incision on her side and tummy to see if she was spayed, but there wasn't any. He suggested I wait and see if she goes into heat. She hasn't shown any signs of heat, but she does have a couple of annoying habits. Periodically she avoids using her litter box and leaves her business on the hall rug. The vet did not find any worms or signs of cystitis. What do you think is wrong with her?

Her indiscriminate urinating and defecating may be symptomatic of the unspayed syndrome. Evidently she is still intact, and although she doesn't display the obvious signs of heat, "displaced littering" is her signal.

Is an unspayed cat apt to have attacks of cystitis?

Yes, quite frequently, because an unspayed female is under a greater amount of emotional stress. When she becomes anxious, it is possible for the effects of her anxiety to attack her bladder, if her bladder is her stress target—her most vulnerable part. Diet and general stress can also contribute.

Will the attacks continue after she is spayed?

Generally, when the primary source of stress is removed, a female is less prone to cystitis attacks.

I have an older spayed female and a seven-month-old female. They have been very close until the past few weeks. Lately, the older one has been hissing at the younger. What is wrong?

The older one has sensed the younger one's coming of age.

You mean she went into silent heat?

Yes, silent to you, but your older cat can sense the high energy level in her young friend. It makes her tense, and so she prefers to avoid her.

So what do I do?

Make an appointment to get the younger one spayed before their tense relationship becomes a habit.

chapter fifteen

sex and breeding: his

I've always had female cats, but now I'm planning to adopt a male kitten. Is it necessary to get him altered?

Yes, indeed. The surgery should be done when he has reached sexual maturity, which can be as early as five and a half months or not until he is around one year old.

How can I tell when he's ready?

Don't worry, you'll recognize the signs. Usually a male's personality takes on a very macho, aggressive flavor.

What do you mean?

He will play rougher, howl at any open window or door, run around like a bandit, maybe puff up his tail, and flatten his ears.

Actually, that doesn't sound too bad. Is he also affected physically?

Yes, his urine will take on a very strong smell. It is difficult to remove the smell of a tomcat from your possessions once he has been spraying around.

I thought any male cat was called a tom.

No, a tomcat is the generic name for an unaltered male, as a queen is the name for an unspayed female.

Aside from the smell, what is all this talk I have heard about spraying?

When a male cat reaches sexual maturity, he frequently christens any object he pleases to mark his territory. When he feels the urge, he will back himself up against the object of his choice, his tail will shoot up in the air, and a spray of urine will emerge. Because of his high sexual energy, he will spray frequently, and this activity will not be limited to his litter box, no matter how well trained the cat may be.

Is it possible to get the smell out of things?

The smell is very persistent, and even after several cleanings, your cat can usually still scent it, which means he may pick the spot to spray again. A strong, pleasant-smelling shampoo is sometimes the answer; or a strong cleaner like Lysol may do it. However, to prevent your male from spraying, it is best to have him altered when his urine and tomcat behavior begin to be offensive.

Suppose a male's urine and behavior never become offensive. Is it still necessary to have him neutered?

I'd recommend it, especially if he will be around other intact cats. In that case the stimulation of the other cats could bring about his sexual maturity.

And then what?

An outdoor, sexually intact cat is subject to stressful feuds. He will often fight to defend his territory from another tom or feud over a female in heat.

Is that why I have seen street cats that look so battered?

Yes, a street tom's life is one long battle—a constant quest for food, shelter, and sex. Often a street tom lives for less than one year. And if a tom is able to hang in there for a while, he is usually old before his time, because of the severe emotional and physical stress.

But if I have my male altered and then let him go outdoors, will he be able to stand up to the toms?

He won't have that macho "knock-'em-down" instinct like the toms, but since he is altered (neutered), he won't be competing with them for females in heat. Also, his sexual energy level will be relaxed, and he won't be driven to brawl to release that energy. But he will still have plenty of energy to defend himself if threatened.

I have a friend who has a two-year-old tom that goes outdoors. He is forever coming home with battle scars, but they don't seem to bother him. My friend takes a lot of pride in his cat's bravado.

Your friend will find, as his tom gets older, that his ability to conquer will decrease as he meets up with younger toms.

I don't mind if my male cat gets into brawls with other toms. To me it is just part of being a cat. Are there any other consequences?

Yes. Think of all the females he is apt to leave with a family; there aren't enough homes now for our kitten population. Why add to the problem?

But I thought it was easy to find homes for kittens!

It is usually easier to find homes for kittens than it is for older cats, but still, many people adopt a kitten on a whim and afterward have a change of heart. Sadly, kittens that are born on the street rarely get even a chance to be adopted.

Isn't there an operation that a male cat can have so that he can still fool around but can't impregnate any unsuspecting females?

Yes, the operation is called a vasectomy.

What does it consist of, and how does it differ from a castration?

The male's testicles are removed in a castration. In a vasectomy the tube that carries the sperm from the testes is severed. While a vasectomy does prevent conception, it doesn't rid a cat of the vital hormones that urge him to spray. So after a vasectomy a male cat still has all his macho tomcat characteristics, except the ability to impregnate.

Can a male cat still spray after he is altered?

It takes about two weeks after the surgery for a male's hormones to change. After that his spraying should stop and his urine should lose that distinctive pungent odor. Actually, a male cat, even after he is altered, is able to spray; but since his hormone level has changed, it will be rare that he does. If he does, the odor is not usually overpowering like that of a tomcat's urine.

Why would an altered male spray?

If he got worked up and excited, he might release his feelings that way.

My neighbor's cat sprayed for months after he was altered, and the smell was still as obnoxious as before the surgery. The cat is now living in the country where it doesn't matter. How can you explain this?

Most likely, when your neighbor's cat was altered, the job was incomplete! When a male cat continues to spray long after he has been altered, it is usually because only one of the male's testicles has been removed.

How could that happen?

Sometimes one of the testicles doesn't descend. The condition is called cryptorchidism. The veterinarian may have removed one testicle that had descended, but didn't search out and remove the other, since that would have made the operation much more complex.

So something could still be done for my neighbor's ex-cat?

Yes, an exploratory operation could be done to locate and remove the missing testicle.

Don't you think the vet would be wrong not to tell my neighbor the real story?

Absolutely! Sending home a supposedly castrated cat with an intact testicle was gross negligence on the veterinarian's part.

Why do you think he would do it?

He might have felt incapable of performing the surgery to remove the undescended testicle, or perhaps he didn't want to take the extra time. However, he should have informed your neighbor that her cat would still be a tomcat for as long as he had the retained testicle. Fortunately, this isn't a common occurrence.

Is it possible for a male and female still to fool around after they are neutered?

Yes, there are cats who carry on as if they were still intact, and some affectionate pairs "act out" the sex act, with the male mounting the female, but with no grand results!

Can an altered male satisfy an unspayed female?

Generally, no. The male is apt to end up with a case of cystitis, and the female with a case of unrequited passion.

What about a spayed female and a tomcat?

A spayed female usually won't tolerate the high energy level and demands of a tomcat. "Paws off" is her feeling!

I have a four-year-old altered male named Celery. Last week a tomcat came into our yard and wandered all over our property and onto the porch. Fortunately, Celery was inside the house, but he watched from the window and hissed away. Later that day, when Celery went outside, he sprayed like a tomcat in several spots on the porch. That evening he strained a bit in his litter box when he urinated, but by the next morning he was fine. What happened to him?

Celery was excited and stressed by the tomcat's high sexual energy, so he left his "trademark" on the porch to mark his territory, especially where the tomcat had left his scent. Celery's anxiety affected his bladder, so he strained when urinating in his box. Fortunately this was only a mild spasm and not a full-blown case of cystitis.

What do I do if the tomcat appears again?

Persuade him to leave by giving him a spritz with the plant sprayer. If you can locate his person or family, try to suggest diplomatically that it would be advantageous to have him altered.

While I persuade the tomcat to leave, what is the best thing to do with Celery?

Distract him and offer him food in a room where he can't view the tomcat.

Suppose they have a run-in outside?

If you are around when it happens, use water—a bucketful, if necessary—to separate them from each other, Whatever you do, don't try to pick Celery up or get in the midst of their brawl. You might become injured—a victim of displaced aggression. (See pages 140–141, How to Treat Your Cat Bite.)

Have you ever encountered chronic victims of the "tomcat syndrome"?

Indeed I have! The most dramatic case was my own cat Sunny-Blue, at that time quite young.

What was he like? How did you come to get him, and what did you do to help him?

Okay, I'll take it from the top. Sam, our sixteen-year-old Siamese, had died. He had been ailing for a while, and Baggins, his nine-year-old companion, had accepted Sam's limited energy level long before Sam reached the end. However, in spite of Sam's ailments, he and Baggins had a beautiful relationship. Baggins cuddled, embraced, and groomed Sam up until his last day.

It seemed best to let Baggins be the only cat for a while after Sam's passing. Because he had sensed Sam's "end" and adjusted to it long before we did, Baggins was not totally bereft without Sam. Two months passed, and we felt it was time for Baggins to have a new companion. While I thought Baggins could adapt to a young or adolescent cat, we weren't ready to cope with the high energy level of a kitten, so we decided to give a home to a cat who was less in demand than a kitten.

Further, Baggins was used to the scent, voice, and touch of a Siamese, and so were we. Therefore we decided to adopt whatever Siamese cat came along—providing it wasn't too mature to adjust to Baggins.

We put the word out among our friends, and for a while nothing materialized. But one evening, Maria, a nurse at The Cat Practice, called to tell us that a woman uptown had found a male Siamese on the street. As the woman already had several cats, plus a few up for adoption, this newcomer needed a home quickly. I called the woman and told her we would pick him up the next day. When she exclaimed, "But don't you even want to see him first?" I answered that I was sure he would be fine.

So what happened?

We drove to the woman's apartment and, while I waited in the car, Paul went in to pick up Baggins's new friend. I had a gut feeling that the cat would be fine for us, and I was sure Paul would instinctively know if he wasn't. Anyway, Paul returned with Sunny's head sticking out of the pillowcase we had brought along to put him in. While Paul drove, Sunny sat regally and calmly on my lap. He definitely suited the mood of our 1953 Bentley.

How old was Sunny?

We estimated his age to be eighteen months—a full eighteen months of pure tomcat. We couldn't bring a tomcat home to Baggins, so Paul took him straight to The Cat Practice and neutered him. It would have been best to wait then for a couple of weeks while his hormone levels diminished before we introduced him to Baggins, but I decided that I could handle the situation if there were any dilemmas.

Were there?

Yes, in spite of the fact that Sunny came home tranquilized. My plan was for Paul to emerge from the elevator and deposit Sunny in the hallway. I would open our apartment door, and Baggins would stroll out and encounter Sunny. As the hall was neutral territory, Baggins wouldn't feel that he had to protect his domain. However, when Paul emerged from the elevator, and before Baggins ventured out the door, Sunny bounded into our apartment like a streak and tackled the unsuspecting Baggins. So we used the plant mister to cool them off, separated them, and gave Sunny another dose of anti-anxiety drug.

Did things get better?

No—in fact, the plot thickened! Paul, along with Baggins, became a victim of Sunny's tomcat syndrome. Because Sunny's tomcat characteristics had been a part of his personality for a good while, any source of high energy threatened him, and he attacked its source.

What did that have to do with Paul?

Paul's energy level was higher than mine. His body language and movements were not so soft. Sunny, a tomcat already threatened by Baggins, transferred his anxiety to Paul as well, so Paul became another one of his targets! All Paul had to do was move quickly, raise his voice, or make a sudden movement of his body, and Sunny would spring after him. Any direct look from Paul would threaten Sunny. He would flatten his ears and prepare for attack. Several times he tried to wrap himself around Paul's arm or leg and position himself for a bite.

How did you deal with this uptight cat?

Sunny's dosage was increased, but primarily we kept a very low energy level—we had no visitors to excite him, we moved about very slowly, and we also started Baggins on the tranquilizer to reduce his angst from Sunny.

Why Baggins?

Because Baggins was wary of Sunny, for good reason, and the drug relieved his anticipation anxiety. With both Sunny and Baggins being treated, the tension decreased.

You haven't mentioned Sunny's personality when he wasn't anxious.

When Sunny was good, he was very, very good, but when Sunny was mad, he was horrid! When he was relaxed and calm, he was exquisitely responsive to contact, but if overstimulated or excited, he behaved like a bandit. (See Chapter 5, Emotions, and Chapter 10, Catsonality Problems, for similar problems.)

What was the outcome?

As Sunny's tomcat hormone levels decreased, his stress tolerance increased, helped by our positive support and the auxiliary support of the tranquilizer.

How did you get Baggins and Sunny to relate well to each other?

Well, we had tried to use my method of introduction (Chapter 6, Choice and Introduction of a Companion for Your Cat). The tranquilizer helped a lot to relieve their tension, but it was butter that actually did the trick!

Sunny liked to lick a lot—whatever he could get his tongue on. One afternoon I increased both cats' dosage of the drug and smeared butter over Baggins's back. Next, I lured Sunny over to Baggins: I let him smell the butter on my fingers, and slowly, as Sunny followed me, I moved closer to Baggins. Then, of course, Sunny discovered butter on Baggins's back.

But what did Baggins do when Sunny licked him?

Baggins was so mellow from the drug by then that he was able to accept Sunny's licks without feeling threatened. I distracted Sunny before he licked too long, because I knew he would get overexcited and then ornery in spite of the medication.

After Sunny had been licking Baggins for a few days, I put butter on Sunny, and Baggins returned the favor. He had been used to grooming Sam, so licking Sunny was a continuation of an old habit.

Why did you decide to use butter?

Both cats enjoyed licking it off our fingers, and I knew it would be hard for them to resist.

Paul must have been very happy to see Baggins and Sunny get reconciled.

Yes, indeed, Paul was ecstatic. He had found it hard to screen his behavior in his own home.

And things were okay between Sunny and Baggins?

Delightful! They still had their spats, of course, but they were everyday cat spats. Sunny wasn't out to "kill" as he was before.

How did it feel to follow your own advice in bringing Baggins together with Sunny?

It was damned difficult! But I knew that it could be done, and that it was up to me to do it. Also, I knew that Paul would, in spite of Sunny's reaction to him, give me all the support he possibly could.

How did things turn out between Sunny and Paul?

They were thick as glue! But Sunny still played rougher and tougher with Paul than he did with me, because Paul's energy brought out Sunny's aggressiveness. However, since Sunny was no longer a tomcat, even his toughness was reasonable.

Do many tomcats act as hostile as Sunny did?

There are different degrees of reaction; everything depends on the tomcat's personality structure, his environment, and other related circumstances.

Sunny is a prime example. His personality structure exudes high, bubbling energy, his reflexes are like Zorro's, his physique is small but well-muscled, and he can vanish like Houdini. Combine all these high-powered characteristics with the fury and tension gained on the street—Sunny's stress tolerance was pushed beyond its limits.

How do you think he ended up on the street?

He might have been an outdoor cat who strayed from home lured by the scent of a female in heat or some other attraction. But there is a strong possibility that his people didn't know enough to have him altered and abandoned him because they couldn't cope with his behavior.

So Sunny's basic personality didn't really change after he was altered?

Precisely! He is still basically the same cat but without the tomcat syndrome; he is less anxious, his stress tolerance has increased, he is more comfortable with himself, and, although he can be incorrigible, he is now compatible with Baggins and with us.

Why did you choose the name Sunny-Blue for him?

Because the word Sunny is a wonderful word to say. It has such a warm and peaceful connotation that when I say it, I am filled with warmth and happiness. This feeling is transferred to Sunny whenever we speak his name. The exception is when we call out his name "in vain," after he has created a heap of mischief! Furthermore, Sunny adores being warm and frequently basks in the sun or any toasty spot he can find. The Blue stands for his gorgeous blue eyes.

Weren't you taking an incredible chance when you adopted a tomcat to be Baggins's companion?

Yes, indeed, but I felt we could work it out. I wouldn't recommend it.

How long were your cats on a tranquilizer?

For about two months. Then, as their stress tolerance increased, their need for the tranquilizer decreased, so I slowly tapered their dosage until they were off it.

Could you have used a homeopathic remedy or Chinese herbs to ease the aggression and Baggins's anticipation of fear?

Yes, they are definitely options.

You mentioned that an intact male is quite vulnerable to cystitis. Is he commonly affected by any other physical problems?

Quiet frequently an intact male becomes obstructed. Whereas cystitis is an inflammation in the bladder, an obstruction means that the urinary passage is completely blocked and the cat is unable to urinate.

You mean he can't go at all?

That's right—he may strain and strain but produce, if anything, only a few drops. If he is not given help very quickly—time is of the essence—the obstruction will be fatal.

Why, what will happen?

The bladder swells as far as it can, and the pressure causes the kidneys to shut down, causing a condition known as uremia.

What is the treatment for obstruction?

Generally, the cat has to stay at the hospital, where the vet will relieve the obstruction and treat the cat for the uremia. The cat's condition can be critical for days. Usually the vet will recommend castration.

So, after he is altered, there is no chance of future urinary attacks?

The percentage of occurrence is less. However, once a male's bladder is sensitized, it becomes more vulnerable. Then, when he is under a lot of stress, his bladder may react adversely. But attacks will occur less often.

What happens if an altered cat keeps getting obstructed?

He can be treated with an operation called a urethrostomy. It is a surgical procedure that widens the urinary passage and eases the urine flow.

It sounds very painful.

There is some discomfort for the cat immediately following the surgery, but the fact that the patient can urinate without any difficulty is enough compensation. With the operation chances are greater than 90 percent that he will not have another obstruction.

Does his diet or activity have to be curtailed?

You must take the same precautions as for any cat with cystitis. Remember, the surgical procedure just gives him a wider opening. (See section Daily Care for Cystitis, pages 142–143.)

chapter sixteen

senior cats—the sixth sense is love

Your cat's five senses—hearing, sight, taste, touch, and smell—have always been so keen. But now that he has marked his tenth birthday, you notice he is not as alert as he was. His ears no longer perk up at the smallest sound, and his appetite has fallen off.

There are measures you can take to help a cat cope with the failings of age. Your love and understanding can provide your cat with the sixth sense he needs to make his daily life easier. Here are some pointers:

- Hearing: If your cat's hearing starts to fail, raise your voice when you talk to him. Don't take him by surprise if he doesn't see you coming; instead, move more loudly. You can also announce your appearance with a knock on an object next to him. He will feel the vibration.
- Sight: Your cat's sight may decline. If so, refrain from redecorating or rearranging your furniture. Your cat knows his way around by furniture position. Any changes will disturb his sense of navigation and frustrate him. If you must make changes, walk him through them.
- Taste: Food may be less appealing if your cat's taste buds get rusty. Tempt him with his preferred treats, varied menus, and a taste of your own food if he is interested.

- Smell: Your cat's appetite is greatly influenced by his ability to smell. Spice up his meals with potent tidbits—a bit of fishy broth or a morsel of sardine—or heat them up in the microwave if his ability to smell wavers. If he can't smell his food, his taste will be impaired.
- Touch: There was a time when your cat adored rough-housing. You could pick him up and wrap him around your shoulders, and he would love it. But lately he complains if you are not gentle with him. He avoids his favorite shelf at the top of the closet, and he really seems to have slowed down. Your cat's limbs may not be as limber as they were, and he no longer has the pep he had before.

Give him a thorough rubdown with your hands to help his circulation. Fix him a sunny spot to snooze in or provide a heating pad set at a low temperature; heat is soothing for his body. Try to anticipate his needs and carry him to his destination without making him feel undignified.

Your love will support and enhance your cat's five senses. Give the sixth sense all you can to make your cat's days happy and comfortable.

If my cat is seven years old, doesn't that make him sixty-three in people years?

Many people think that a cat older than six is an old cat; and to complicate things, they multiply each cat year by nine. Perhaps this common myth has something to do with the idea that a cat has nine lives for every one of ours. However, this way of figuring a cat's age is incorrect and confusing. In fact, I wouldn't consider any cat a senior cat until he was over ten years old. The average life span of a cat is usually about sixteen years. Don't make your cat geriatric before his time.

I have a twelve-year-old cat who is in very good shape, but I have noticed she drinks a lot of water. Should I be concerned?

I would have her urine and blood checked by the vet, to make sure her kidneys are all right. Kidney problems are a common old-age malady for cats. Fluid therapy of lactated Ringer's can be given under the skin to wash out the toxins. It's a lifesaver for many a cat.

What do you recommend to compensate for kidney degeneration?

Carbohydrates are a necessary source of nutrition for older cats. Try giving your cat creamed soups, noodles, and even a taste of cookies and cake, if your cat is at all interested. Consult your vet for kidney-friendly foods.

I have noticed that my thirteen-year-old Jimmy seems to be a little stiff when he walks and stretches out. How can I help him?

Sounds like Jimmy's joints are getting a little tight. Groom him daily and massage his body—both actions will stimulate his circulation and help to ease his crinks. Fix him a cozy spot in the sun, as heat is soothing. If he is sedentary, encourage him to take little jaunts. He might welcome your company.

My seventeen-year-old Molly just can't get enough of our laps. She always has to be sprawled out on us. What do you make of it?

Molly probably seeks the heat from your body, since it warms up her joints. Also, she enjoys having close, constant contact with you now that she is getting along in years.

Does a cat's fur change color with age, and what about the texture and thickness of the coat?

Sometimes dark-colored cats such as black or Siamese get touches of white or gray in their fur as they mature. Generally the texture of the fur doesn't change, nor does their fur thin or thicken.

My thirteen-year-old cat Jenny has a chronic constipation problem. I add bran to her food and give her the medication and laxatives that the vet prescribed, but even so the vet often has to anesthetize Jenny to clean her out. Can you think of anything else?

Constipation is another common problem with age. You might talk to your vet about starting Jenny on a tranquilizer to relax her body. Sometimes fluid therapy can help to bring relief.

Why a tranquilizer?

Because a tranquilizer will help to relax Jenny's body. The more she relaxes, the less her body contracts and the more it can expand inside. If she can breathe more easily, the supply of oxygen to her body will aid her circulation and ultimately help her to move things through her digestive system better. Also, the tranquilizer will prevent her from becoming frightened and tense when she tries to have a bowel movement. That is important, because her previous bowel movements have been difficult, and she is wary of the whole process now.

Have you heard of any other cases like Jenny's?

Joyce, a former editor of mine, had a cat, Christopher, who was eleven years old and experienced severe bouts of constipation. In spite of a regulated diet that included raw liver, bran, and natural grain laxatives, and a daily enema (given with a baby's enema bulb), Christopher still became constipated and had to be anesthetized weekly so he could be relieved. Joyce became especially concerned because Christopher literally shook and shivered right up to the vet's cleaning-out procedure, and it took him a couple of days to calm down afterward. Joyce hoped I might suggest a remedy for Christopher's chronic condition.

I explained to Joyce that Christopher's emotional reaction to the anesthesia, plus her concern and worry, intensified his condition. When he became upset, he breathed with difficulty, and his body contracted, increasing his inability to have an easy bowel movement. Christopher needed a tranquilizer to relieve his anxiety and relax his muscles. Valium, I suggested, can usually produce both results effectively.

Joyce could understand that Christopher was affected by the procedure, but she felt he was very calm and placid generally and that a tranquilizer might make him too tranquil.

Although Christopher might be outwardly calm, I told Joyce, he invariably internalized his tension, and the stress affected his bowel movements. A tranquilizer would give his body the relaxation he needed to help him get it all out. I also told Joyce to be sure to give Christopher extra love and comfort to keep him in a happy state of mind. Joyce decided to give it a try.

What was the outcome?

At first Christopher was wobbly with the tranquilizer, but within a couple of days he was able to void regularly (although he still required a special diet and, less frequently, warm-water enemas), he actively sought more attention, and he appeared to have a happy glow. He began to spend more time with his companion, Topper, whom he had lately been avoiding. Best of all, Joyce was able to cancel his next appointment to be cleaned out, whereas before then his daily home regimen usually had become ineffective within a couple of weeks and he had needed the extra boost. Eventually his daily medication as well as his enemas was decreased, and an anesthetized cleanout became an exception rather than a rule.

My cat, Spike, is fourteen years old, and although he acts like a teenager, I think his eyesight is failing him. What can I do?

I would have his eyes checked by the vet to see if medication or anything else can help him. In addition, don't rearrange your furniture or change your furnishings, because if Spike has difficulty, he counts on his memory of where objects are to keep him from bumping into things and avoid confusion.

Rambo is an eighteen-year-old cat who lives up to his "powerful" name. Although he is blind, he flicks his tail and ripples his back if he feels his guardians are too helpful. He still calls the shots with his companion Lucy. She looks the other way when he walks out of line.

What can I anticipate to make my cat's life more comfortable as he approaches old age?

You will notice that your cat may not be able to do things as easily and gracefully as he did before. Maturity and impairment of a cat's five senses usually go hand in hand. However, don't call attention to your cat's inability and insult his dignity by laughing at him or scolding him if he breaks something. Instead, try to make things easier for him so he doesn't become tired out and frustrated.

What are some examples of making things easier for him?

If he has trouble climbing up, you might provide him with step stools. Carry him into bed at night if he sleeps with you. Help him get to a high place if he desires to go there, and help him down again. (Refer to "Anticipate Your Cat's Senior Years" in this chapter.)

My two cats were very close until Charlie, who is three years older than the other one, became sick. Now the younger one spends most of his time outside, and sometimes he picks on Charlie. Why does he do this?

He picks on Charlie because he is frustrated and doesn't understand why Charlie isn't his old self. Also, a cat's natural instinct is to protect himself from sickness and danger by withdrawing from the source— in this case, Charlie.

Recently our middle cat, Macho, died of a lingering illness. Since then we have noticed that our senior cat, Piper, who is fourteen, is more relaxed and appears to be lighter in both body and spirit. He is even closer with Tucket, his two-year-old companion. Do you think it is because he felt the stress from Macho's condition?

Quite frequently a cat can be a "medical detective" and sense a companion's sickness before it actually surfaces or is diagnosed. Piper probably felt the stress and discomfort of Macho's condition, and it affected his day-to-day living. With Macho at peace, Piper's anxiety level has eased, and he even can cope with his young and energetic companion.

Is it usual for a cat to behave strangely during his companion's sickness?

Yes, it is quite common. Brutus is an example. When his companion, Bridget, a ten-year-old, suddenly had a breathing attack, he ran to her and put his paw across her chest. When their person, Marsha, went to pick up Bridget, Brutus did not want to let Marsha near her. Finally Marsha was able to distract Brutus. As she carefully picked up Bridget and prepared to carry her out the apartment door, Brutus ran into the hall, which he usually

feared, and howled. Marsha had to distract him again so she could rush Bridget to the vet.

Bridget never came home from the vet's. It was a critical heart problem. Evidently Brutus sensed how sick his Bridget was and didn't want to leave her side.

Are heart conditions a big concern with cats?

Heart problems can seem to come up quite suddenly, especially in senior cats. One evening a client called to give me an update on her two cats. As we talked, she heard a cry from her senior cat Rafaello. He was crouched on the rug and his breath was labored. I asked for the number of her emergency vet so I could call ahead. It was not a long car ride, but Rafaello didn't make it. The vet's feeling was that he had had a heart attack—perhaps cardiomyopathy, which can sometimes go undetected.

Is it common for a senior cat to withdraw from his companion(s)?

Yes, especially if the senior cat has a physical problem. He may not be able to tolerate too much activity, and it is easier for him to cope if he simply withdraws.

It is hard to see your cat unable to participate in life and enjoy it as he once could, but you can help him grow old gracefully with loving attention and understanding of his needs.

My cat has a critical illness, and I know he will not last long. What can I do to make things easier for him, and how will I know when it is his time to go?

If he is responsive to contact, hold and stroke him frequently. Talk to him whenever you are close by. The more you reach out to him, the better he will feel. Carry him into bed at night if it is hard for him to navigate. Encourage him to eat by fixing him his favorite foods. Hold his plate for him or hand-feed him, so he will feel your positive energy.

Our two younger cats used to have trouble getting along with each other, and our cat Timmy, who was then twelve years old, was their security blanket. Now Timmy is sixteen, and he has developed a few chronic health problems. Timmy has little appetite and is very gloomy. What really worries us is that he used to be such good friends with both Amber and Perri, his companions, but now he hisses whenever they approach him and stays off by himself. What do you make of this, and what can we do to make Timmy feel better?

It sounds like Timmy cannot deal with Amber and Perri's high energy level. He protects himself by being unfriendly to them and withdrawing. To relieve Timmy's anxiety, stimulate his appetite, and help relax his skeletal muscles, you might start him on a small dosage of an anti-anxiety drug. (Consult your vet for dosage instructions.) Otherwise, it sounds as though you are very loving to Timmy and sympathetic to his needs.

Do you find that older cats become less tolerant of their people with age?

Yes, this is a fairly common characteristic of senior cats. Their stress tolerance can often decrease along with their physical attributes. They may lash out at their people and companions in order to protect themselves.

Exactly what do you mean?

Lucky, a fourteen-year-old male, was a good example of such a cat. His person was at her wit's end when she called me; she and her husband were desperate with worry about Lucky. They had been referred to me by Paul, who had seen Lucky at the hospital where he did surgery.

She told me that Lucky had always been a moody cat, but in the past month he had attacked her neighbor and daughter with a vengeance, biting and scratching them severely. They wanted to help Lucky and themselves before he had another outburst. Someone had advised them to have his teeth and claws removed to prevent any further injuries, but they didn't feel they could condone such a procedure.

Fortunately, my house call to Lucky was a success. After collecting his case history and observing Lucky as he lay stretched out by the bathroom door, I could tell that Lucky preferred to expend the least energy possible merely to move his body around. His facial expression was tense, and his breathing was a bit labored. Paul had started him on medication to make his body more relaxed and to treat an internal inflammation.

I explained that Lucky's physical condition had lowered his stress tolerance and made him a perfect target for aggressive outbursts. Whenever he was overly stimulated—by too much noise, sudden movements, or too many people around—he was apt to lash out at the source of anxiety. However, they could prevent Lucky's striking out if they could protect him from provoking stimuli. For example, when their daughter had her playmates over, Lucky should be kept isolated from them to prevent his becoming excited.

If they were able to interpret Lucky's "hands off" signals, such as a ripple of the back or lash of the tail, they would know when he was anxious and leave him to himself. (Refer to the chart in Chapter 5, Emotions.) I added that an anti-anxiety drug would help solve Lucky's problem.

Lucky's person seemed to get a good grasp of what I told her. I felt that Lucky was in good hands and that his family would stand less chance of being victimized by his attacks now that his problem was understood.

My next conversation with Lucky's person was a positive one: Lucky was calmer, and the family was no longer living in fear of a sudden attack. A few months later I received a note from his person. Lucky's physical ailments had gradually increased, and he had died peacefully. They were very grateful that they had been able to take care of Lucky and make him comfortable during his last days.

A cat's life is dependent on his person's love and support, his own physical strength, his inner will or spirit, and fate. If one of these elements is weak, the others can support him until it is reinvigorated. But if the missing element is vital and cannot be restored, the cat's end is near.

You can tell that your cat is near the end if he has a faraway, withdrawn look in his eyes, is unresponsive to contact, and hides. If at this time he also appears to be in pain, it is time to have your vet assist him to end his life. It is best to have him reach the end before he reaches unbearable pain.

Harpo is a splendid fourteen-year-old cat whose person, Bobby, has helped him to live his on into his senior years. Harpo formerly lived in

Bobby's restaurant, but when Harpo's appetite and weight declined and a physical examination indicated that he had a mass in his abdomen, Bobby took him home to join his other cats. At first it appeared that the mass would have to be removed surgically, but I felt that Harpo might not survive the emotional and physical stress from the operation. Bobby wanted whatever Harpo needed to live a happy live. Paul started Harpo on medication, and with Bobby's moral and loving support, Harpo rallied!

Angel, another senior cat, is fifteen years old and formerly lived in a sweet shop. When the shop closed, Angel moved in with her person's sister. It was a major transition to which Angel adapted with flying colors. True, she doesn't have constant visitors, but she has a real live person to sleep with every night.

Mabel, another shop cat, was the inspiration and namesake for Mabel's Specialty Boutique, a cottage-like shop that sold handmade items—influenced by many animals, primarily cats. Mabel eventually passed on from cancer at age thirteen, but her softness and refinement lived on in the shop's ambiance and objects. The trademark also bore her image.

Sometimes a cat dies suddenly without giving any kind of major distress signals. This type of loss can be very difficult, because there is no time to accept or prepare for it.

I am reminded of my late cat Baggins. When Baggins died, his only apparent problem was his teeth. An inflammation in his mouth didn't respond to medication, so Paul took a blood sample from Baggins. The results were poor and indicated a severe illness. We tried giving him medication.

The next afternoon Baggins showed no interest in lunch, which was quite unlike him. Then I noticed that his breathing was labored. Hourly, his condition grew worse. I sat up with him most of the night, so I could be sure he wasn't in pain. By morning Baggins was no better. With each breath, he became more rigid and uncomfortable. We had no choice but to let Baggins go. Although he was just nine years old, it was his time. He and Sunny-Blue had just started to really snuggle up close to each other, the way Baggins and Sam had done. Sam had passed on the year before, but perhaps

his need for Baggins was stronger than ours, and Baggins had to answer his call.

What is the best way to dispose of a cat's body after death?

You can make arrangements through your veterinarian to have the cat buried at an animal cemetery or cremated. If you live outside the city where there is some space, you may be able to bury the cat yourself.

How can I make provisions for my cat's life in case he outlives me?

Have your lawyer make provisions for your cat in your will, or arrange for a friend or relative to be responsible for him.

My twelve-year-old male, Inky, passed on recently. Would you advise adopting a cat for his companion, who is ten years old?

If your cat appears lonely, by all means adopt a friend for him. If he is in good health and peppy, adopt a spunky kitten. But if he tends to be reserved, a mellow kitten would be the best choice. Adopt a male if your cat had a good relationship with Inky.

Our two-year-old cat's companion passed on several weeks ago. We are wondering if we should adopt a kitten for her. So far, she doesn't appear despondent, and we have been giving her a lot of attention. She has been with us less than a year.

Perhaps you should wait and get to know each other better. When she is ready for a friend, you will be able to tell by her behavior. She will become restless, hard to please, and possibly destructive. Or she may sleep a great deal and often appear bored.

We had had Sunny-Blue less than a year when our cat Baggins passed on. Sunny had quite an insecure past, and because we hadn't had much time together and he lost Baggins so abruptly, we delayed the adoption of a new friend until his relationship with us was more secure. Perhaps the three of us needed more time to release the mournful feelings we still had for Baggins.

Anticipate Your Cat's Senior Years

As your cat approaches his tenth year, you can ease him into his senior years by providing the following care:

1. Have him checked out by the vet every six months.
2. A urinalysis every few months will indicate if his kidneys are in good shape.
3. There are special foods available for the senior cat. You can, also, add brewer's yeast and wheat germ to his food, as they help him cope with daily stress. But if there's a kidney problem, avoid brewer's yeast, because it is high in protein and may be deleterious to the kidneys.
4. Stimulate the cat's circulation by daily grooming.
5. Salt his food so that he will drink more water, which helps to flush out his kidneys. Filtered, bottled, or distilled water is best.
6. Be aware of your cat's weight. It is best that he be slightly chubby, in case he goes off his food.
7. Carbohydrate treats are a good source of energy in case of kidney degeneration. (Try him with noodles, creamed food, rice.)
8. Build his self-esteem by complimenting him. The happier he is, the better he will feel. Contact is very important. Hug and pet him frequently. Ignore his protests. Insist, and he'll respond.

Coping with Loss When the Time Comes

When the end comes near for your cat, remind yourself of the wonderful and loving times you shared together. Although death is as natural as life, it is not as easy for most people to accept. However, the less we deny our loss, the less painful it will be to accept the inevitable. The following steps will help you prepare for the loss of your cat as painlessly and naturally as possible.

1. If your cat has a long-term chronic illness, he may rally for a while and then decline. A cat in critical condition may not make it. If your cat's physical condition cannot be regulated even with medication, his spirit usually will not be enough to keep him going.

2. Your cat's survival is dependent upon whether his illness can be reversed, whether he has a will to live, and your love and support. If his condition is critical and one of these elements is lacking, his chance of survival is decreased.
3. If your cat's prognosis is poor but he appears comfortable, happy, and not in pain, there's no reason to consider the end. You don't have to have him put down. As long as your cat is content and receptive to touch and can sustain his dignity, there's every reason to continue his life. He may even reach the end naturally. But if your cat begins to withdraw or hide, displays apparent discomfort, and appears visibly un-catlike, it's time to seek the veterinarian's help, so your cat can depart easily and gracefully without undue or prolonged pain
4. It's not uncommon for a cat to stop eating almost completely but still to be high-spirited and alert. If so, he can be maintained for a while with fluids to keep him hydrated. If he seems depressed, you don't want to maintain him artificially. But as long as he has the desire to go on, give him the support he needs.
5. Frequently you can tell that a cat is near the end when he gets a very faraway look in his eyes and is devoid of appetite, no longer in contact, withdrawn, undesirous of your affection and may hide. At this point you need to be in touch with the veterinarian.
6. The more observant you can be of your cat's behavior, the more able you will be to help him reach his end with grace and ease. If a cat has to be given anesthesia, it will take effect quickly and easily if the cat is ready to go.
7. If your veterinarian needs to assist your cat when it's his time to go, you'll know then whether you want to be present. If you feel you can comfort your cat and not go to pieces, it may be the best answer for both of you.

When I was at The Cat Practice, I devised ways to provide the ultimate in support and care to the client and patient in coping with the end. Dr. Carol Fudin, whose cats were patients of ours, and who is now a well-known grief counselor in Manhattan, modeled her classes for veterinary practitioners on my concepts. Dr. Fudin gives supportive counseling to those who find it difficult to cope after their cat has passed on. Many veterinary clinics now offer such counseling, and the Animal Medical Center in Manhattan has a program of counseling sessions.

8. Try to decide in advance what to do with your cat's body. You might consult your veterinarian for information about burial or cremation. Have your attorney prepare a will with your cats included as beneficiaries in case you die first, or make an agreement with a friend or relative to look after your cats.

   An old and dear friend of mine always maintained that his white cat, named Kitty Cat, had been the single most constant female in his life. He passed on quite prematurely and hadn't made any provisions for Kitty Cat and her companion, Grey. Fortunately, the right home for them materialized.

9. Because a cat is a natural medium for fluctuations in surrounding energy fields, both human and animal (I refer to this ability as "cat sense"), he can often sense an internal, organic change before it's externally evident. That's why it isn't uncommon for a cat suddenly, without provocation, to attack or reject a close companion because he senses the companion is sick. He becomes overwhelmed, confused, and/or threatened by his companion's discomfort. The companion becomes a source of anxiety, triggering this erratic behavior. The cat has assumed the role of medical detective.

10. You may find that your former cat's companion is overwhelmed with grief and loneliness. He may need a new friend. On the other hand, if the surviving cat is a senior with chronic health problems, a new cat might be disastrous. Try to comfort the bereaved cat with extra loving care and attention.

It's not uncommon for a surviving cat to exhibit changed behavior. J. Audrey was a nine-year-old cat who lived for years with Muggsy, a four-year-old male, and an older female. J. Audrey had always been quite feminine and demure. She was indeed third on the cat pole, but when the older female died J. Audrey emerged from the cat closet. Perhaps she had been overawed by her companions' dominant catsonalities and had opted to keep a low

profile. Now, her self-esteem increased and her assertive, extroverted behavior was indicative of this reintegration. She now only had one dominant cat to deal with, so Muggsy soon learned to share his life with a liberated kitty!

Easing Your Own Bereavement Overload

1. Plan your time so you don't have empty lapses. It's only natural to mourn the loss of a dear friend, but your life must go on in a healthy way. If you formerly spent noontime each day with your cat, schedule a new activity that makes you feel good. You can think of your late cat, but don't try to hold time in place.
2. If your cat passed on suddenly without warning, your grief may be devastating, but think of the cat as being at peace.
3. Let go of guilt that's frequently unjustified. Your cat loved you, too, and you gave him a good life. It is a tribute to him if you continue to love and nurture cats and other creatures.

It is not unheard of for people to wish to dedicate a memorial to their late beloved cat. One of my clients, a physician from Georgetown, arranged with her clergyman to present a service in memory of her sixteen-year-old cat, Rug. Another client gave a monetary gift to The Humane Society of New York, where I'm a consultant, to initiate a new program in tribute to his late senior cat, Gunge. Even if only a token, it can help.

It's important to remember that cats, as well as dogs, horses, elephants, cows, apes, and others of the animal kingdom, including us human beings, are only here for a time. While on this earth, all deserve to be treated with concern, care and respect, and accepted with delight for what they are. I know that cats definitely add much to the substance of our lives, and in innumerable ways.

A cat is a sociably responsible being that offers us unconditional love. A cat does not judge, but is tolerant and forgiving. Once you have been duly chastised for a transgression (purely as a point of instruction, of course), a cat can let go and move on. What a powerful example!

A cat's power of concentration is unsurpassed. When a cat wants something, his or her focus and patience is unlimited. I'm reminded of a cat spread out before a mouse hole; or one that lies quietly nearby, listening intently to your changing breath sounds as you awaken, until you are ready to rise and take up the daily routine together.

A cat is also a multitasker, seducing you with his body, outwitting you with his mind and embracing you with his spirit. Is it any wonder why we can never get enough?

We can laugh with glee at the frisky antics of a tiny kitten, marvel during times of sorrow at the deep concern and understanding evident in the steady gaze of a cat and, in the evenings, sit quietly and let our breathing slow to match the rumbling purr of a tired but contented companion, enjoying the light stroking of a patch of rumpled fur.

In some very important ways, cats are very much alike. But they are not interchangeable. If you have been a loyal guardian and protector to only one cat over the years, you may cherish memories of sharing your space with that special creature so much so that you can't imagine bringing any other into your home. It's understandable for you to feel that way.

Yet, if you have lived with more than one cat, either singly or in family groupings, you have surely learned to appreciate that cats have many different catsonalities. And, although you may try to resist, their cute behaviors, myriad quirks and, yes, even cattitude will win you over, so that each individual soon finds its own particular place in your heart. That's when you come to recognize that all cats are special. As guardian to such a creature, you should count yourself honored—and special as well.

So, if a cat deigns to live with you, rejoice, for whether you know it now or not, you have been found worthy of a great gift. The opportunity to observe, approach, and get to know a magical creature that, although wild in ways and manner, is content at this time to also be your friend and companion.

How-To Quick Reference Index

General Index

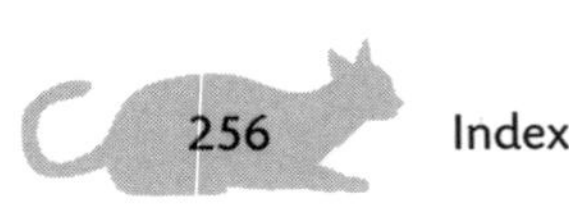

Photo by Sheril McCormack

About the Author

CAROLE C. WILBOURN quickly and easily answers to the title The Cat Therapist. In 1973, she co-founded, with Dr. Paul Rowan, a hospital devoted to cats. Over several years there, her intensive work with cats having all kinds of physical and often psychological problems led to her understanding of the necessity for treating the "whole cat." Concerned with the emotional well-being of her patients, Carole developed and currently specializes in adapting her "Wilbourn Way" cat treatment programs to cats the world over with psychological and emotional disorders, treating her patients through "house calls" or phone consultations with their guardians.

Longtime consultant to The Humane Society of New York, Carole Wilbourn wrote a monthly column for *Cat Fancy* magazine for sixteen years, and has written several books on cats. She is currently in residence at Westside Veterinary Center in Manhattan, seeing clients there by appointment and working with its hospital patients. Carole also writes a column for *New York Tales* magazine, serves as online columnist for "In Defense of Animals" (www.idausa.org) and schedules media appearances, lectures, library events—and writing time for an upcoming book when she can.

Carole Wilbourn lives in New York City with, of course, a cat—a Rescued Siamese male named Orion, after the constellation. Orion's down-to-earth guardian can be reached through her website thecattherapist.com.